Vodou in Haitian Life and Culture

Vodou in Haitian Life and Culture

Invisible Powers

Claudine Michel
and
Patrick Bellegarde-Smith

palgrave
macmillan

VODOU IN HAITIAN LIFE AND CULTURE

First published in 2006 by
PALGRAVE MACMILLAN™
175 Fifth Avenue, New York, N.Y. 10010 and
Houndmills, Basingstoke, Hampshire, England RG21 6XS
Companies and representatives throughout the world.

PALGRAVE MACMILLAN is the global academic imprint of the Palgrave Macmillan division of St. Martin's Press, LLC and of Palgrave Macmillan Ltd. Macmillan® is a registered trademark in the United States, United Kingdom and other countries. Palgrave is a registered trademark in the European Union and other countries.

ISBN 978-1-349-53305-3 ISBN 978-0-312-37620-8 (eBook)
DOI 10.1057/9780312376208

Library of Congress Cataloging-in-Publication Data
Vodou in Haitian life and culture : invisible powers / edited by Claudine Michel and Patrick Bellegarde-Smith
p. cm.
Includes bibliographical references and index.
ISBN 978-1-4039-7161-6 (alk. paper)—
1. Voodooism. 2. Haiti—Religious life and customs. I. Michel, Claudine. II. Bellegarde-Smith, Patrick.

BL2490. V62 2006
299.6′75097294—dc22 2006043152

A catalogue record for this book is available from the British Library.

Design by Newgen Imaging Systems (P) Ltd., Chennai, India.

First edition: October 2006

10 9 8 7 6 5 4 3 2 1

Transferred to Digital Printing in 2008

To the 13 original founding members of KOSANBA who in 1997
had the vision and the courage to create the first international
scholarly association for the study of Haitian Vodou
and
To the *manbo, houngan, hounsi*, and other Vodouists who
believed enough in the *Lwa* to allow us to capture
their encounters with *les invisibles*

Hymon T. Johnson: "What can Christianity learn from other religions?"
Archbishop Desmond Tutu: "That God is not a Christian."
Santa Barbara, California, November 4, 2005

Published under the auspices of *The Congress of Santa Barbara* (KOSANBA),
a scholarly association for the study of Haitian Vodou, and the
Center for Black Studies, University of California, Santa Barbara

Contents

Invisible Powers: An Introduction

Invisible powers can be made visible. They can be made visible by action, acts by humans on the great stage of life. In Vodou, the *sevitè*, women and men who serve the spirits by literally embodying the divine *Lwa, les invisibles*, manifest themselves during *sèvis*, the ceremonies and rituals in which trances occur. During the *sèvis*, the invisible and visible interact with surprising intimacy. Invisible powers can also be illuminated through research. This volume presents the work of prominent scholars in the field of Vodou studies who offer their expansive views of a religious system which has generally been either unseen or misperceived.

For most of the past five centuries, Western civilization has deliberately demonized peoples of African descent as an easy justification for their enslavement. Africans were considered to be less than human. Their physical features were declared repulsively ugly. Their cultures, denigrated. Their religions? Nonexistent, or a compendium of heteroclite, ill-conceived notions of noxious superstitions emanating from pre-literate and pre-scientific peoples who never quite rose from practicalities into the rarefied realms of abstract thinking. As logic, sophisticated science, languages, and religion became the apanage of the West, African religions were dismissed with terms such as polytheism, primitivism, paganism, heathenism, and animism, seen through European eyes as impediments to progress and material development. The patronization that informed the "white man's burden" became a liberal notion whereby the little brown brother might be educated and elevated beyond his primitive beliefs. And why not try? Many brown brothers and sisters fell into the trap, abandoning their genetic and cultural inheritances.

This predjudice is still common currency in American discourse. "Voodoo economics" or "voodoo politics" are part of a political arsenal in which "black" magic defines the Other from American goodness and munificence. Today, Western powers continue to meddle in others' affairs via government, the private sector, and through missionary workers acting as "agents of civilization." Hollywood, the film industry, and the media perpetuate negative stereotypes. The United States and—by extension—other Caribbean and Latin American republics, justify the military occupation of Haiti using the same "white man's burden" principles. As slaves were denied their full humanity, Haiti is denied its sovereignty.

Generations of educated Haitians, taught to speak and write in French, were also taught to embrace the ideals of their imperialist neighbors and the logic of colonial or neocolonial power relationships, individually and collectively deprecating Haiti, its citizens and its unique culture. The chapter by Carrol F. Coates is particularly significant in this regard. He studied a half dozen novelists who, though none admitted to practicing Vodou, have generally given a positive spin to Vodou. One author in particular proudly confessed to interviewing *houngan*, priests, in his effort at verisimilitude. The real story is the increasing acceptance of Vodou by middle- and upper-class Haitians. These writers constitute a "who's who" of progressive Haitian politics; their lives parallel anticolonial struggles around the world.

Patrick Bellegarde-Smith extends the foregoing analysis by arguing that socioeconomic development, to be secure and genuine, must always take into account the national culture. The modernization of cultural elements—the acceptance of the Haitian language, Kreyol, and the Vodou religion—are necessary conditions for the realization of balanced development, befitting an autonomous culture and an independent country. Neither democracy nor development can be "spread" from a beneficent West to others; this is merely the white man's burden revisited. In both Coates and Bellegarde-Smith, one sees that social elites have realized that the emperor (imperialism) has, in fact, no clothes. One is reminded of an African proverb: "run from a naked man offering you a piece of cloth."

Claudine Michel provides a clear and succinct explanation of Vodou as superstructure. Vodou transcends its religious role, becoming a spiritual discipline that infuses all other societal systems. Gerdès Fleurant addresses the "song of freedom" and the impact of Vodou music from the country's genesis to the present day, and in the musicians' search for renewal through modernizing their art. As a defining element par excellence, these two chapters show how Vodou provides an integrated and integrative worldview/worldsense that has an effect on every aspect of the Vodouist's life: family structure, economics, healing, and so on. This understanding goes a long way in explicating Haiti's development.

Karen McCarthy Brown, author of the modern classic *Mama Lola*, presents Haiti as a case study in Afro-Caribbean spirituality. Her chapter is a precise and eloquent description and analysis of the religion, with an in-depth discussion of its healing powers—healing in all its permutations, defined broadly. This latter theme is augmented by Pierre Minn in his ongoing research interest on Vodou within his areas of specialization: illness, healing, medicine, and related nomenclature. Largely descriptive, Minn's chapter is an excellent introduction to that field.

The text by Richard Brent Turner reminds us of the Haitian cultural elements found in New Orleans, Louisiana. While Haiti alone cannot be credited with "Hoodoo"—as the origins of U.S. Blacks parallel those of Haitians—aspects of Haitian Vodou, as well as African religions, were absorbed into the rituals of Christianity practiced by American Creoles. In New Orleans and beyond, something of the African *religionwissenschaft* resisted conversion to Protestantism and retained strong Africanisms in its practice of both Catholicism and contemporary Vodou as currently encountered in that part of the world. In Elizabeth McAlister's chapter, one finds the lingering and predictable colonial influences in the adoption of forms of anti-Semitism in the Haitian *rara* festivals around Christian Easter. McAlister addresses "the demonization by European Christianity of two groups—Jews and Black Africans." But, she states, things are never as simple as they seem, since even the Other has agency—a recurring leitmotiv found throughout this volume.

Last but not least are the detailed explanations of specific elements of the religion within overarching contexts of Vodou as worldview. Leslie G. Desmangles reminds one that the Christian cross has alternate African meanings found in the Vodou cross and the *vèvè*. It completes our understanding that culture need not be enslaved to materiality; enslaved Africans who crossed the Atlantic with no possessions, not even clothing—"*yon men devan, yon men derye*," one hand aft, one fore to hide their nakedness—carried their culture with them. Similarly, LeGrace Benson, in her pioneering work on Islamic influences on Haitian Vodou, speaks for the many thousands of enslaved Moslems who made it across and whose impact on Haitian religious culture remains unheralded. Anna Wexler considers sequined Vodou flags as religious art and their role in rituals and culture. Donald Cosentino's chapter is an *envoi*, un *bonswa-dam*, the grande finale, which illustrates through a major *Lwa*—Sen Jak Maje—how the invisible is rendered visible in Haitian metaphysics, how Vodou followers interact with spirits with astounding familiarity, and how this invisible world impacts the daily existence of Vodouists.

The editors selected texts they felt were of great significance in presenting an image or images of Vodou in a number of different disciplinary perspectives. Each of the authors we chose has made serious contributions to the growing field of Vodou studies. Together, these scholars account for hundreds of years, many generations of research, and one hopes that the accumulated wisdom is commensurate. And none make apologies for wanting to rectify false information that passes for knowledge. All would admit to having fallen to Haiti's *envoûtement*, to being enchanted by Haiti.

This book could not have been completed without the support and commitment of many individuals and institutions. First, we thank our contributors for their willingness to be part of this project and acknowledge the publishers that graciously allowed us to reprint some rare material. These are as follows: The Brooking Institution, Macmillan, Oxford University Press, The UCLA Fowler Museum, *Sociological Analysis*, and *The Journal of Haitian Studies*. We are grateful to the anonymous reviewers who read this manuscript and offered constructive comments. We are indebted to our editor at Palgrave Macmillan, Gabriella Pearce, who was enthusiastic about this project from its inception and provided strong encouragement and invaluable editorial direction. We could not have asked for a more supportive and visionary editor. We thank Professor Oyèrónkẹ́ Oyěwùmí for telling us that Palgrave would be a great press for this project. We also acknowledge the support and patience of various members of the publication staff at Palgrave Macmillan, editorial assistant Joanna Mericle, the designer, for the beautiful design; and, of course, painter Hërsza Barjon for her superb rendition of Gede Nibo who graces and blesses the cover.

The Center for Black Studies at the University of California, Santa Barbara, houses both the Haitian Studies Association's *Journal of Haitian Studies* and the Congress of Santa Barbara (KOSANBA), a scholarly association for the study of Haitian Vodou; we are grateful for the support offered to both associations and the researchers who have contributed to this anthology. Members of the Center's administrative and editorial staff have been incredibly supportive of our work and initiatives. We are beholden to three core members of the Center: Chryss Yost, our publication manager, Amber Wallace, managing editor, and Mahsheed Ayoub, business officer, who have assisted once again with administrative and research support and multiple editorial tasks. We also thank the Office of Research and the College of Letters and Science at UCSB for their support of the research conducted at the Center for Black Studies and in the Department of Black Studies, in particular Michael Witherell, Vice Chancellor of Research, and Melvin Oliver, Dean of Social Sciences. Professors James D. Smith, Anna Everett, Christopher McAuley and Julie Carlson must be acknowledged as well as Louise Moore, Nancy Doner.

Our profound gratitude to our mother, sisters, and friends, Jacqueline Epingle, Myrtha St-Louis, Minou and Mamoune Placide, Paulette Denis, Nirva Chérasard for stoic performances during ceremonies to Nanny Placide for her incredibly elaborate tables for the *sèvis* for the *Lwa*; and, always, to our cherished Mama Lola, a pioneer in the defense of Vodou, for opening her home and sharing with us her knowledge and wisdom and her family. Final thanks go to Claudine's husband, Douglas H. Daniels and daughter, Kyrah M. Daniels, for their support of KOSANBA and for participating and helping with field research activities and conferences and, for generously allowing us to have the space and time to work on *one more* Haiti project.

Claudine Michel and
Patrick Bellegarde-Smith

Chapter 1

Afro-Caribbean Spirituality: A Haitian Case Study

Karen McCarthy Brown

Haitian views of healing and wholeness as revealed in the religious system called Vodou provide the focus for this study. While the specifics of the discussion would differ if it were centered in other Caribbean locales, there are certain basic attitudes and understandings about the nature of the human condition and the causes and cures of human suffering that are broadly shared among descendants of African slaves throughout the Caribbean—areas that may be collectively named the Afro-Caribbean. Before turning to Haiti, I will first consider briefly the factors that create the differences among Afro-Caribbean cultures and then attempt to outline the common foundation on which their various healing systems rest.

Traditional attitudes and practices surrounding health and spirituality vary from one area of the Caribbean to another for several reasons. Of first-level importance is the place (or places) in Africa from which the slave populations were drawn and the resulting ideas about health and spirituality that the slaves brought with them. For example, in Haiti there are three clear lines of African influence: those of the Fon peoples, most of whom live in the area we now call Benin; the Yoruba peoples (Nigeria); and the Kongo peoples (Angola and Bas-Zaire). By contrast Cuban traditional religion is dominated by Yoruba influence, while that of Jamaica has its deepest roots among the Akan of Ghana. Other factors that account for the differences are the nature of the slave systems under which the first generations labored, including the brand of Christianity practiced by the slaveholders; the geography, plant and animal life of the New World setting and the differences and similarities that the slaves found between these and the ecologies of their homelands; and the social, political and economic history subsequent to slavery.

In relation to Haiti, the last point warrants special comment. Haiti was the second independent republic in the Western Hemisphere and the first Black one. After its successful slave revolution (1791–1804) and mainly as the result of trade boycotts,

Haiti was effectively cut off from contact with the United States and Europe for nearly a century. Furthermore, even though the French colonists had established Catholicism as the official religion of the people of the island, including its slave population, Haiti was denied priests by the church for more than fifty years following the revolution. At the opening of the nineteenth century, when the long struggle for independence ended, it is possible that as many as three-quarters of the slave population of Haiti had been born in Africa. Therefore, for a substantial period of time following the expulsion of the French, several strong African cultural traditions interacted in Haiti in an atmosphere relatively free of outside influence. This phenomenon sharply distinguishes Haiti from the rest of the Caribbean and particularly from places such as Jamaica. Jamaica experienced a continuing colonial presence well into the twentieth century. As a result, the influences of Africa in Jamaican traditional spirituality are subtler and more diffuse than those in Haiti. However, the ubiquitous "balm yards" or healing centers in contemporary Jamaica are significant African survivals. It is likely that their existence is a testament to the durability of a level of religious practice that does not require elaborate temples or rituals, or the participation of large numbers of persons. More importantly, their survival is also evidence of the centrality of healing for African-based spirituality.

In spite of diverse input from Africa and divergent experiences during and after the period of slavery, the various Afro-Caribbean communities share a broad range of traditional assumptions, attitudes, and practices relating to health and healing. I have identified six such factors, which I believe to be common to the healing traditions of the Afro-Caribbean.

First, healing is the *primary* business of these religious systems. In fact, it is not an overstatement to say that spirituality and healing are synonymous in the Afro-Caribbean. Client-practitioner interactions occasioned by problems in the lives of individual persons occupy much of the time of spiritual leaders. Furthermore, even large ritual events that occur on a regular basis can be understood as healing ceremonies when placed in their proper context.

Second, the understanding of personhood operative within these Afro-Caribbean healing traditions is a fundamentally relational one. The individual person is defined by a web of relationships that includes not only the extended family but also the ancestors and the spirits or saints. Furthermore, the individual *qua* individual is also understood in relational terms. Personhood is seen as constituted by a dynamic balance of diverse spiritual energies or tendencies.

Third, healing within Afro-Caribbean traditions takes place through ritual adjustments in these relational webs. To be more specific, healing involves adjusting or reactivating the reciprocal gift-giving that characterizes all relationships in the Afro-Caribbean, whether they are relationships with the living, the dead, or the divine.

Fourth, these African-based religious traditions address a wide variety of maladies. The expertise of the healer extends beyond physical problems to include social problems arising from such areas as love, work, and family life. While a person with physical symptoms could well be given herbal treatment appropriate to those symptoms, herbs would not represent the main part of the cure. In fact, the distinction between physical and social maladies is finally an insignificant one. Basic diagnostic categories are concerned with the *origins* of problems, and problems are virtually always seen as due to disruptions in relationships. The major curative action is therefore, as we have seen, directed at healing relationships. Further, the connection between a specific cause (the root problem) and a particular set of symptoms (the presenting problem) is by no means a necessary one. In other words, failure to honor the spirits could equally well result in the loss of a job or in digestive difficulties.

Fifth, these healing systems have a penchant for working through what Lévi-Strauss called "the science of the concrete."[1] The harmful emotional states that cause disruptions in relationships—such as jealousy, despair, fear, anger—are addressed in ways that appeal to the nonrational and even nonverbal dimensions of human interaction. Emotional or relational states are concretized in sounds, gestures, or objects that are laden with the highly condensed metaphoric referents of such things as taste, smell, and color. Adjustments are then made in the externalized or concretized relational situation. For example, in Haiti, a marriage threatened by the destructive anger of the husband could be treated by placing ice and a little sugar syrup in a jar that also contains a slip of paper with his name written on it several times. The jar is then inverted, the basic signal within the Vodou science of the concrete that a situation is to be changed. The wife, who desires to "cool down" and "sweeten" her husband, "works the point" several times a day. She lights a candle by the jar, prays over it, and concentrates her energy on the desired end. Scientific and social-scientific thinkers alike have tended to label this sort of healing practice "magic" or "superstition," thus dismissing it from the larger psychotherapeutic discussion, where it could well suggest middle-range alternatives to drug therapy and the talking cure.

Finally, all of these traditions are involved in one stage or another of negotiation with Great Atlantic culture, that is, with the Western world. Scientific medicine, capitalism, individualism, and modern technology all present challenges to customary attitudes and practices in the area of health. In some parts of the Caribbean, exposure to these forces has been substantial and long-term (Puerto Rico, for example), and as a result, traditional healers have circumscribed their activity, focusing on problems that would be considered insignificant by the church and by Western medical institutions, such as broken love affairs, predictive dreams, and chains of bad luck. By contrast, in rural Haiti the majority of problems of all sorts are still treated by traditional healers. Yet no area in the Caribbean has been without some contact with the trappings of modern life. African-based systems of spiritual healing characteristically accommodate elements of modernity in their worldview rather than react to them competitively or with hostility. For example, a traditional healer may advise a patient to go to a hospital or get a shot of penicillin from the local clinic. Unfortunately, there has not been the same openness in the other direction.

This summary view of the Caribbean context serves as background to a more detailed discussion of traditional healing in Haiti, which will begin with sections on the centrality of family and the view of person. The focus on exchange relationships emerging from these two topics will provide the organizing motif for discussions of Vodou rituals and of the Vodou spirits. A more specific treatment of the Haitian Vodou understandings of the causes and cures of human suffering will follow. This will touch on a variety of topics, including the etiology of problems, the sources of authority used in treatment, and the questions of morality that arise in the quest for healing.

Serving the Spirits in Haiti

Haitians do not often call their religion "Vodou," a term that in the rural areas at least is still reserved for a particular subtype of dance and ritualizing. (*Vodou* comes from the Fon language and means "spirit.") When Haitians refer to the religious dimension of their lives they refer to a form of activity rather than an institutional entity. They say they "serve the spirits." I have come to believe that human suffering is the major impetus for serving the spirits and, furthermore, that an understanding of Vodou ritualizing in terms of the ways in which it both comprehends suffering and ameliorates suffering yields greater insight than any other.

"Moun fèt pou mouri" (People are born to die), Haitians are fond of saying with a shrug of the shoulders. This proverb comments on the suffering and death that are commonplace occurrences in poverty-stricken Haiti and shows the stoic acceptance that, on one level at least, characterizes the Haitian attitude toward such a life. Haitians have no vision of heaven in their religion,[2] no ideology of progress shaping their understanding of history, and virtually no experience of upward mobility in their lives or the lives of their children. Suffering is an expected, recurrent condition. It is not an exaggeration to say that problem-free periods in life are pervaded with an anxiety that anticipates crisis just around the corner. Life as a whole is thus characterized by cycles of luck and the absence of luck. The clever, faithful, and/or powerful person is one who manages by a juggling of scarce resources to give generously to the living, the dead, and the spirits. The resulting network of dependents who are obliged to serve and elders or social superiors who are obliged to give sustenance and protection—even though subject to the inherent unpredictability of personal relationships—provides the only means any Haitian has of controlling his or her "luck." At the very least, the obligations created by these gifts construct the safety net that is essential for survival, given the uncertainties of life in Haiti.

Haiti occupies the western third of Hispaniola, an island it shares with the Dominican Republic. It is a small country—about the size of the state of Maryland—that is home to 5.5 to 6 million people. Haiti is still largely an agricultural country, although much of its land has been rendered nearly useless by short-range farming techniques and soil erosion caused by cutting trees to produce the charcoal most people still use to cook their food. Diseases such as tuberculosis, malaria, yaws, syphilis, and elephantiasis, which have been virtually eliminated in most of the Western Hemisphere, afflict the population in Haiti still. In parts of Haiti the infant mortality rate is above 50 percent, and anyone reaching the age of fifty-five or sixty is considered among the fortunate. The majority of children show some signs of malnutrition: spindly arms and legs, swollen bellies, reddish brittle hair. Social disease is also rampant in Haiti, a country that has survived a succession of brutal dictators who have increased their personal wealth at the expense of the people and maintained their power through random violence and intimidation. It is estimated that 80 percent of the population is illiterate and that the average annual income for a Haitian is somewhere between $200 and $300. When the considerable wealth of the 8 to 9 percent of the population known as "the elite" is taken into account, it appears most persons in Haiti get by on little more than $100 a year—and yet a chicken purchased in Port-au-Prince can cost as much as $5.

"Mizè mennen parespè," the Haitians say, meaning, if you show you are suffering people lose respect for you. *Mizè* (literally, "misery") is an interesting word choice here, for while it can be used to refer to suffering in general, it is used most often to refer to poverty with all its attendant pains and indignities. There are many beggars in Haiti. One sees them everywhere, but most often in markets, cemeteries, and churchyards. In spite of their numbers, there is a special shame associated with begging. This becomes apparent in the way begging is used within the Vodou system. When the spirits want to teach a lesson in humility to a devotee, they command that person to don the ritual version of rags and go to the market and beg. The ignominy of begging comes largely from the fact that beggars are seen as isolated individuals whose activity announces to the world that they have been abandoned by the extended kin group and now must forage on their own. Even if the family were lost through death rather than discord, the person who must beg can easily be seen as someone who was not clever enough or respectful enough or sufficiently hardworking to find a place as adopted kin in another family.

The Centrality of Family

For the slaves taken from Africa, the loss of extended family was so great that they apparently made efforts to recreate that family before they had even set foot on the shores of the New World. It is reported that some slaves recognized an incest prohibition as existing between males and females who had undergone the Middle Passage on the same ship. We know almost nothing about the interactions among slaves in the early part of the eighteenth century, when large numbers of them arrived in Haiti to work the plantations. However, knowledge of the crucial role of the extended family throughout West Africa easily leads to the conclusion that whatever blending among Fon, Yoruba, and Kongo cultures took place during that period must have been compelled in large part by the need for family. In the early stages this need would have been met through fictive kinship structures in which putative "mothers" and "fathers," "aunts" and "uncles," and "cousins" provided the individual with both identity (a place in society) and protection. Since the contributing African cultures defined family as including the ancestors and the spirits, the need for family was both a social and a spiritual need.

The slaves' loss of access to family land in Africa was as great as their loss of the African family itself. Indeed, from one perspective family and land were inseparable. Prevented from visiting family graves and from leaving food offerings and pouring libations at ancestral shrines, the enslaved African had also been denied the means of ensuring the spiritual blessing and protection of the ancestors. Thus when slaves could bring no other possessions with them, some nevertheless managed to carry away small sacks of the soil of their motherland. This connection of family, land, and religion persists in rural Haiti today.

Unlike most of the other Caribbean nations Haiti is predominantly a country of peasant farmers, many of whom own their own land. Where the social structures have not been decimated by the combined pressures of overpopulation, depleted soil, and corrupt politics, rural people in Haiti tend to live in large, patriarchal, extended families. Even moderately successful men in the countryside may enter into multiple *plasaj* or common-law unions with women. Each of these women is set up in a house of her own in which she raises the children born of their union. Thus a multigenerational extended family can swell to considerable size even when counting only the blood kin. Such families, however, are not defined solely by blood ties. Large rural families invariably include adopted "godmothers," "godfathers," and "cousins," as well as a number of "maids" and other workers who exchange their labor for a place to sleep and for meager rations. Included in this latter group are the *restavèk* (literally, the "stay-withs"), children whose parents could not afford to feed them and so either sold them or gave them away to slightly more prosperous families. Social hierarchy is relentless in Haiti. There is always someone poorer than oneself. Even the most minimal rural household with only one or two able-bodied adults to work an unproductive square of earth manages to have a servant.

The patriarch of the extended family functions as the *oungan* or priest when that family serves the spirits. He is often the one who treats family members when they become ill, although an outsider may be called in for such treatments if there is someone in the vicinity who has a reputation as one who "knows leaves." However, it is necessarily the patriarch who presides at the *gwo sèvis*, the big dancing and drumming events that include animal sacrifice. These ceremonies are held annually if family resources permit. More commonly they are held at longer intervals and then only in response to crises within the group. The purpose of the elaborate ritualizing is to honor, entertain, and feed the ancestors and the Vodou spirits which those ancestors served.

The family dead are buried on the family land and the cemetery is a major center for religious activity. In addition, a cult house for the ritual objects of the family is often built on a small, separate plot of land. Thus, to inherit land is also to inherit the bones of the ancestors and the duty to honor those ancestors as well as to serve the spirits represented in the cult house. Conversely, to be separated from the land is also to risk one's access to the power and protection that these spirit entities provide.

Separation from land and family is, however, an increasingly frequent experience for the younger generations of Haiti's rural poor. Inheritance laws in Haiti work to divide the land into smaller and smaller plots. This pressure, combined with that of the multiple problems cited earlier, has pushed large numbers of young people off the land and toward the elusive promise of a better life in the cities.

For young men urban life is often cruel. In the countryside they are reared to the expectations of male privilege and power. (Even the female-headed households that are prevalent in the cities perpetuate this ideology to a degree.) Yet some experts estimate unemployment among young urban males at 60 percent and others argue that the figure should be much higher. Women fare somewhat better in the urban environment. Most of the factory jobs available are of the piecework variety, and European and American employers seem to favor women for these repetitive tasks. Urban women also have a market tradition bequeathed to them by their rural sisters.

In the country it is the women who take the excess produce to market, along with bread, candy, herbal teas, baskets, and other things they make with their own hands. The urban woman spun away from the rural extended family frequently ends up not only in charge of her house and her children—as she might well have been in the country—but also solely responsible for their financial support. In the countryside her market money would have been the "rainy-day savings" for times of drought and hunger or the means to fulfill a private dream for herself or her children. In the cities she must rely on the old market skills more centrally. The poor urban woman is constantly engaged in small-scale commerce, often in several such enterprises simultaneously. For example, even if she has a regular job, she may sell peanut candy at the door of her home or work as a seamstress or beautician in the evenings and on weekends.

Both men and women who no longer live with their extended families feel the loss acutely. In fact, this sense of loss can persist for generations. In the cities, it is the Vodou temple and the fictive kinship network it provides that compensates for the missing large rural family. The head of the temple is called "mother" or "father," and the initiates are known as "children of the house." The Vodou initiate owes service and loyalty to his or her Vodou parent after the pattern of filial piety owed all parents by their children in Haiti. In turn, Vodou parents, like actual ones, owe their children protection, care, and help in times of trouble. In certain circumstances this help is of a very tangible sort: food, a place to sleep, assistance in finding work. The urban Vodou temples are currently the closest thing to a social welfare system that exists in Haiti.

The differences between men's and women's lives in the cities have also left their mark on the practice of urban Vodou. While in some parts of rural Haiti women can gain recognition and prestige as *manbo* (priestesses), herbalists, or *fanm saj* (midwives), nowhere in the countryside do they effectively challenge the spiritual hegemony of the male. This is not the case in the cities, where there are probably as many women as men in positions of religious leadership.

The urban Vodou temples run by men tend to mimic the patriarchal structure of the rural extended families. The urban *oungan* is notorious for fathering many children and recruiting desirable young women to be among his *ounsi*, brides of the gods, the ritual chorus and general workforce of a Vodou temple.[3] He thus creates a highly visible father role which he then operates out of in relation to all those who serve the spirits under his tutelage. While the female *manbo* who heads a temple is not necessarily more democratic in all of her relationships with those that serve the spirits in her house, she does tend to be so in the ways that a mother's role is normally less authoritarian than that of a father. For example, many temples headed by women function as day-care centers for the working mothers associated with them. In sum, the woman-headed temple tends to reiterate the tone and atmosphere inside the home, a place where women have usually been in charge. This is an atmosphere that allows for more flexibility in human relationships than is found in the male-headed temple, which recalls the more public and therefore more rigid social rules of the entire extended family. This shift toward greater authority for women in urban Vodou has undoubtedly had an effect on the nature of the care given to individuals who turn to traditional religion to solve the many problems that urban life in Haiti can bring.

Whether the temple is headed by a man or a woman, it is clear that its appeal to the urban population is rooted in its ability to recreate family. A song sung at the beginning of Vodou ceremonies in Port-au-Prince illustrates this:

Lafanmi semble,
Semble nan.
Se Kreyòl nou yè,
Pa genyen Gine enkò.

The family is assembled,
Gathered in.
We are Creoles,
Who have Africa no longer.

The Vodou View of Person

In Vodou, persons are said to possess several "souls." In fact, there is no generic term in the Haitian Creole language that includes all of these spiritual entities or energies, even though each possesses some of the characteristics of what Westerners call soul. Furthermore, the word *nam*, derivative of the French word for soul, is only one of the complex of forces that constitute a person. A person's *nam* is usually understood as the animating force of the body. The most immediate effect of death is the departure of the *nam*, which is sometimes said to linger for a short period of time around the corpse or grave. The *nam* is an evanescent thing that disappears soon after death.

By contrast the *gwo bonanj*, the big guardian angel, is capable of sustained existence apart from the body it inhabits. One of the situations in which the person is separated from his or her *gwo bonanj* occurs during the possession trance, which is central to Vodou ritualizing. The struggle that marks the onset of possession is understood as a struggle between a person's *gwo bonanj* and the Vodou *lwa* (spirit), who desires to "ride" that person and to use his or her body and voice to communicate with the faithful. One who is thus ridden by the spirit is known as a *chwal* (horse) of the spirit. Those who are possessed report that they lose consciousness after this initial struggle. The loss of consciousness and the resulting amnesia about what the spirit said and did while riding the *chwal* is explained as due to the departure of the *gwo bonanj*.

Similarly, it is the *gwo bonanj* that wanders from the body during sleep, even into the land of the dead, thus allowing deceased persons or those living at a great distance to appear in dreams. The wanderings of the big guardian angel during sleep are sometimes useful for information-gathering. For example, a mother in Haiti said she learned from a dream that her daughter in New York had met with an accident and broken her arm. In like fashion, when a person is uneasy, she may say that her *gwo bonanj* is agitated. This is an undesirable state mainly because it robs the person of sound sleep and therefore of dreams, which are an important vehicle for communication with the dispersed family, the ancestors, and the spirits.

Finally, it is the *gwo bonanj* that must be ritually removed "from the head" of a person shortly after death. The big guardian angel is then sent "under the water" to dwell for a period of time until it (now referred to as a *mò*, one of the dead) is "called up from the water," installed in a clay pot known as a *govi*, and placed on the family altar. The Vodou ceremony known as *rele mò nan dlo*, calling the dead from the water, calls them from Gine, Africa, a watery land said to exist below the earth. The ceremony ideally takes place a year and a day following the death. Because it is an elaborate and expensive ceremony, however, in practice families wait until there are several of their dead whom they may retrieve at once. As a result the dead frequently emerge complaining of cold, dampness, and neglect. In this ceremony, the dead speak through a kind of ventriloquism possession and genuinely sound as if they come from both far away and underwater. Their identity is confirmed by the intimate knowledge of family life which they display. The spirits called up from the waters of Africa inquire about family members and comment on problems within the group. Given these various understandings of the nature and activity of the *gwo bonanj*, it seems fair to conclude that this dimension of soul is both the consciousness and essential personality of the individual.

The *ti bonanj* (little guardian angel), which each person also possesses, is much more difficult to define. One urban *manbo*, or priestess, gave me two interesting responses to questions about the nature of the *ti bonanj*. "When you look at your shadow," she said, "you will see that sometimes it has a dark center. That is the *gwo bonanj*, but the paler shadow around the dark center is the *ti bonanj*." When asked what the little guardian angel does, she gave another concrete illustration: "When you are walking a long way or carrying something very heavy and feel so tired that you know you are not going to make it, it is the *ti bonanj* that takes over so you can do what you have to do." The *ti bonanj* is thus perhaps best described as a spiritual reserve tank. It is an energy or presence within the person that is dimmer or deeper than consciousness, but it is nevertheless there to be called upon in situations of stress and depletion.

Much less routinely, Vodou *oungan* and *manbo* speak of another dimension of the person called the *zetwal* or star. This is not an inner presence so much as it is a kind of celestial parallel self. The concept of the *zetwal* is rooted in the belief that each person is born with his or her fate already foreknown and unchangeable. The regular movements of the stars and their recurring patterns mimic, perhaps even direct, the larger contours of life in the human community. Whatever control an individual has over his or her life thus comes in specific moments and short-run situations. *Mizè* (suffering) may be held at bay only for a short time and *chans* (luck) only marginally enhanced. The overall shape and direction of a life are determined by fate.

The *nam*, the *gwo bonanj*, the *ti bonanj*, and the *zetwal* are the constitutive parts of a Haitian view of personhood that is clearly derivative of what ethnographers call the "multiple soul complex" in West Africa. The fact that Vodou contains European elements as well as African is also hinted at in this formulation. In addition to their Catholicism, the French planter class of Haiti was known for its participation in a variety of forms of marginal spirituality including Freemasonry and spiritualism. It seems likely that the astrological flavor of the *zetwal* concept also owes its parentage to this line of influence, even though the notion that individual persons are born with their fate already cast by the gods was a belief held by the Fon and to some extent also by the Yoruba.[4]

While Vodou devotees understand the dead body *(kòr kadav)* of a person to be a material substance separable from these various animating spiritual entities and therefore subject to decay and ultimate dissolution, the body/soul or material/spiritual split is not central to their understanding of personhood. As an indication of this it is worth noting that there is no division within the Vodou view of person between drives or appetites that come from the body—for example, hunger and sexuality—and those that come from the spirit or mind. In fact, sexuality is perhaps the central animating force in all of life. Much of Vodou ritualizing suggests that sexual and spiritual energy come from the same source.

What complicates the understanding of personhood is the realization that individuals are not comprehensible apart from the Vodou spirits associated with them. It is easiest to discuss this in the urban setting, which I know best. Here, each person is said to have a *mèt tet*, master of the head. This is the main spirit served by that person, and if the person is one who serves as a "horse" of the spirits, it will be the *mèt tet* who most often possesses that person. To a certain extent the personality of the individual human being mirrors that of his or her *mèt tet*. For example, a man who has Ogou as his *mèt tet* will be expected to exhibit some of the warrior spirit's anger, strictness, and perseverance in his everyday behavior. Yet he will also have been told that Ogou is "too hot" to be served alone. The spirit of war and anger must be balanced by others, for example, by a strong "sweet" spirit such as the ancient and venerable snake spirit, Dambala.

In addition to a *mèt tet*, each individual has a smaller number of other spirits, usually two or three, from whom he or she receives special protection. This complex of spirits, which may consist of some that are known only in that family and others that are recognized throughout Haiti, differs from individual to individual. It is partly because of this that Vodou, though centrally concerned with morality, could never produce a codified moral law that would apply equally to all persons. In Vodou, an individual lives a moral life by faithfully serving the particular configuration of spirits that "love" or "protect" that person. This includes following their advice, advice that will be consistent with the personalities of the spirits. Thus it might be said that the Vodou ethic is an intensely contextual one.

It is the urban devotee's particular grouping of protective spirits that determines the nature of ritual as well as moral obligations. Furthermore, it is important to note that the choice of this penumbra of protective spirits is not for "the living" to make; Vodou devotees insist that it is the spirits who choose the persons they love or protect. Yet, priests and priestesses do determine the choices the spirits have made, often through divination.

A question may well be raised as to whether the Vodou spirits are truly distinct and separate from the persons who serve them. This question is answered in paradoxical ways within Vodou ritualizing. Beliefs surrounding possession trance and the struggle of the *gwo bonanj* with the possessing spirit, as well as the insistence that the person is chosen by the spirit and not vice versa, point to a clear distinction between spirit and person. However, from the perspective of certain rituals such as those that occur during initiation and after death, the individual person cannot be separated from the spirits that reside "in the head" or "on" the person, these being equally common expressions among Vodou devotees. Initiation rituals simultaneously "feed the spirits in the head" and establish a repository for them outside the person. This repository is called a *pò tet* (head pot). After initiation it is placed on the Vodou family altar and becomes the focus of rituals designed to cool, soothe, and strengthen the person. Furthermore, when the spirit is removed from a person's head at death, this spirit is sometimes treated as if it were the *gwo bonanj* and sometimes as if it were the *lwa*, the Vodou spirit, who was the *mèt tet* of the dead person. Similarly, when the ancestor is called up from the waters and established on the family altar, the spirit is called both by the name of the ancestor and by the name of the *lwa*. For example, reference may be made to "Marie's Ogou" or to "Pierre's Dambala." Thus there is also a sense in which at least the head spirit is identified with the *gwo bonanj*, if not with the individual in a larger sense.

In fifteen years of work on Haitian traditional religion, I have learned that paradoxes of this sort are to be cherished rather than resolved, for it is invariably such paradoxical statements that provide the greatest insight into the religious system we call Vodou. If it is understood that within the Vodou worldview the individual is both a separate self and an inseparable part of a family, then it can be grasped how the spirits who are part of that extended family can be *both* other than and merged with those who serve them.

Rituals of Haitian Vodou

For some individuals, coexistence with their spirits presents no problems; life flows more or less smoothly. It may be the case that someone within their family serves the spirits and this is sufficient to fill the hungry bellies, slake the dry throats, and stroke the wounded pride of the ancestors and the *lwa*. However, if one is not so fortunate and life is not going well—and it often is not in a country such as Haiti—then more is required. Vodou offers a series of ritual steps that escalate the intensity of the individual's involvement with the spirits. Each of these ritual steps is based on an exchange. The person commits to service of one sort or another; in return the spirits proffer relief and protection.

Some problems can be handled with a onetime or at least a short-term commitment to the spirits. This type of commitment could be something as simple as lighting a candle before the image of a spirit, or it could be an elaborate and expensive feast for several spirits, which would include dancing, drumming, and animal sacrifice. Other problems require a more routinized and long-term relationship with one or more spirits. Such life-time commitments vary from "marriage" to a spirit to the decision to become a priest or priestess.[5]

In the process of escalating their commitments to the Vodou spirits, devotees accomplish two related things. First, they gradually increase the strength and stability of their own *gwo bonanj*. For those who move to the upper levels of initiation this means mastering the art of possession trance, which is the art of both letting go of the *gwo bonanj* and bringing it back. Second, devotees gradually increase their control over the Vodou spirits as well. Men and women who advance to the grade of *oungan* and *manbo* do so through a ceremony in which they "take the *asson*."[6] The *asson* is a small, hollow gourd covered with a mesh of glass beads and snake vertebrae. This rattle, which is the emblem of the Vodou priesthood, is not used to make music but to signal key changes in the drum rhythms in a Vodou service, as well as to summon and send away the *lwa*. When a *lwa* tries to seat itself on an inexperienced horse, the struggle between the *gwo bonanj* and the spirit can become violent and even harmful to the horse. It is in situations such as these that the spirit must be sent away. Thus, within limits, Vodou priests and priestesses have power over the spirits. As one Vodou priest put it: "The spirits don't like the *asson*, but they give it to us anyway so we can work with them."

Although it is clear that overall the spirits have far greater powers than do the living, the relationship between devotees and spirits is nevertheless characterized by reciprocity and mutual dependence. The *lwa*, like the ancestors, depend on the living to feed them. Hungry spirits are troublesome and destructive. The living, in turn, depend on the protection and luck that only the spirits can guarantee. This relationship is not unlike the one that exists between parents and children. While the greater power and authority of the parents is unquestioned, parental care in Haiti is not purely altruistic. In the rural areas children work from a young age and their work soon becomes essential to the ongoing family enterprise. Play for children four or five years old often consists of small fetching and carrying tasks; and all over Haiti, the childless person is pitied mainly because there will be no one to take care of that individual in old age. For those reared in monotheistic religious traditions, the notion that the spirits are dependent on their devotees is an especially difficult one to grasp. Yet comprehending this principle is essential, for without understanding that the spirits need the living, it is all too easy to attribute the problems, illnesses, and general harassment that the spirits at times dole out to the living as due to their temperamental, or worse, evil nature.

Vodou is a blend of various African traditions with Catholicism. Although it can be argued that Catholicism has been Africanized in Vodou, and that this is a far truer statement than its reverse, this does not mean that the Catholic Church has no role in the life of the 85 to 90 percent of Haitians who serve the spirits. Pilgrimages to various churches and attendance at Mass are integrated into many complicated Vodou rituals. In addition, the church has taken over the major ceremonies of the life cycle. Birth, where it is ritualized at all, is celebrated through baptism. Also, ideally everyone should go through a First Communion. Pictures from this event are among a family's most treasured possessions. For economic reasons, most Haitians enter *plasaj* (common-law) partnerships rather than have legal marriages. However, where there is a wedding, it is understood that it should be a church wedding. The church also buries the dead, although Vodou rituals are woven in and out of the wake, the entombment (burial is aboveground in Haiti), and the memorial Mass that comes nine days following the death.

Vodou ritual pervades the life of the great majority of Haitian people. For example, candles are lighted and libations poured at countless family altars every day. There are also large ceremonies that have a more social and celebratory air. Among these are the sumptuous feasts for the spirits that occur with some frequency throughout the calendar year at large urban temples. These are a source of entertainment and celebration for curious onlookers and invited guests as well as for the members of that particular Vodou family. Even though all guests may not be offered drinks and plates of food, it is a tradition that the doors of the Vodou temple are closed to no one. Furthermore, the more people present at one of these events, the more chance it has of being a success. The spirits will not come until the crowd is *byen eshofe*, well heated up. When sweat is streaming down the bodies of the drummers and they have found that vast reserve of energy on the other side of fatigue, when their intricate polyrhythms drive the dancers to new heights of grace and spirit, when the voices of the leader of songs and the *ounsi* chorus challenge one another in an ascending spiral of statement and response, that is when the ceremony is *byen eshofe* and that is when the *lwa* will mount their horses and ride.

Spirit Possession

Once the spirit is in charge of the horse, the crescendo of energy stops and people settle in to watch the possession performance. The term "possession performance" is not used here to indicate that there is anything false or contrived about these visits from the spirits. Vodou priest and priestess alike condemn the occasional person in their midst who may *pran poz*, act disingenuously as if possessed. The term is used rather to indicate what has often been noticed about possession in the Vodou temple: it has a theatrical quality. The characters of the major Vodou spirits are well known. Even an outsider such as myself can identify the possessing spirit within moments of its arrival because of certain stereotypical behavior as well as the ritual garb and implements that the spirit requests. However, the Vodou priests and priestesses, the ones usually possessed at these large feasts, improvise freely within the character range of the spirit. Thus a *lwa* not only goes through standard ritual salutations and exhibits certain forms of behavior that are seen virtually every time this spirit possesses someone, but the spirit also addresses particular persons and gives advice about specific problems. The spirits hug, hold, and dance with the devotees. They give ritual blessings and sometimes ritual chastisement, both appropriate to the situation. They sing. They eat. They cry. At these large public events, the Vodou spirits process the problems of the community, fine-tuning human relationships. Sometimes an intimate problem can be whispered into the ear of a sympathetic *lwa*, and the spirit will take the devotee aside for a discreet and private audience. More frequently, these interactions with the spirits become the occasion for an individual's problems to be aired (and healed) in the larger community context.

One specific example of this process will serve to make several points. There was a *oungan* in Carrefour (a town on the coast road south and west of Port-au-Prince) who had a reputation for being a strict and dour disciplinarian in his Vodou family. Because she had angered him, he sent away a woman named Simone, the song leader in his temple, and told her never to return. At a ceremony not long after, this *oungan*, whose name was Cesaire, was possessed by the warrior spirit, Ogou. Ogou arrived in a rage and immediately began to berate Cesaire (the very horse he was riding). Who did Cesaire think he was, Ogou asked, that he could send Simone out of the temple? Simone was one of Ogou's favorites, and besides, it was he, Ogou, who was in charge of the temple, not Cesaire. The gathered faithful were instructed to convey this message to the ill-mannered *oungan* without fail, and then the spirit departed, leaving the body of Cesaire in a crumpled heap on the temple floor. When he had barely regained his senses, the reluctant Cesaire was carried along in a procession of all the temple dignitaries, complete with the brightly colored, sequined banners of the temple, right to the home of Simone. They stood outside and sang Vodou songs of invitation and reconciliation. After much coaxing, Simone agreed to come back to the temple, and, accompanied by the full parade, she was ritually reintegrated into the Vodou family.

This example shows something of the complexity of the possession process in which a *lwa* can chastise, even humiliate, his own horse. Yet, perhaps more significantly, it also shows the key role of the community in the interpretation and application of the wisdom of the spirits. Thus, the public airing of community problems and issues within the Vodou temple is a means of enforcing social sanctions, mobilizing the assistance of the community, and mending broken relationships. It is, in short, a way of healing.

Yet there are vast areas of Vodou ritual that are concerned with healing in a more direct way. These vary from the individual client-practitioner interactions (practices that will be discussed below in a section on the types of caring used in Vodou healing) to the expensive and elaborate cycles of initiation rituals.

Initiation

Vodou initiation ceremonies are never undertaken lightly or routinely. Almost always it is trouble with the spirits, manifesting in problems in the individual's life, that lead a person to undergo initiation. In the temples of the Port-au-Prince area there are four levels of initiation possible. Each level involves a period of seclusion that may vary from three to twenty-one days, and most temples have a small interior room set aside for such purposes. Persons tend to be initiated in small groups. The men and women in these groups become "brothers" and "sisters" in a special way. Above all, they are committed to helping each other with ritual duties. This is the case even when the groups contain individuals who are seeking different grades of initiation. All grades of initiation have public rituals that occur intermittently in the exterior temple dancing area as well as rituals reserved for the already-initiated members of the house that occur within the inner chamber.

The first level of initiation is called the *lave tet* (head-washing) and involves cooling and soothing as well as feeding the spirits in a person's head. The second level is *kanzo*, a word that refers to a rite in which initiates are briefly removed from the initiation chamber in order to undergo a ritual trial. In the semipublic part of the *kanzo* ritual, small, hard dumplings are snatched from boiling pots and pressed into the palm of the left hand and the sole of the left foot of the initiate. When this ceremony is completed, the initiates are told: "Now you are *kwit* [cooked]; no one can eat you," that is to say, no one can do harm to you. They are also admonished: "Never say hot again, say strong!"

The third level is called *sou pwen*, on the point. *Pwen* is a complex, multivocal concept in Haitian Vodou, as it is in Haitian culture in general. Within the general culture, "singing the point" or "sending the point" refers to a socially appropriate means of indirect communication that is especially useful for conveying difficult messages. For example, one young man in Haiti told me this story: he was courting a young woman who came from a family as impoverished as his own. The girl's mother decided that the match offered neither one any chance of advancement, and yet she was loathe to insult her daughter's suitor. So when he visited, she went about her household tasks singing a popular song, the refrain of which was "Dè mèg pa fri," (Two lean [pieces of meat] do not fry). The young man got "the point" and broke off his relationship. In and out of the temples, it is often Vodou songs that are used for the purpose of singing the point. These songs have a sparse, even cryptic quality to them that lends itself to communicating several different, sometimes contradictory, meanings at once. The person who "sends a song" in the Vodou temple, that is, the one who suggests the next song to be sung by the group, is not only following a closely prescribed ritual order in which each important *lwa* is saluted in the proper order with his or her own songs and rhythms, but quite frequently is also sending the point, *pwen*, to a person or group of persons present at the ceremony. Such an observation both reveals the extent to which Vodou ritual intertwines with and comments on the life of the community and suggests a preliminary definition for the troublesome word *pwen*. At a level of abstraction uncharacteristic of the way people who serve the spirits speak, *pwen* may be said to mean the condensation or pith of something. At a concrete, ritual level *pwen* are charms or medicines composed of words, objects, gestures, or some combination of the three. They may be drawn on the earth, spoken, sung over a person, placed under the skin, or ingested; they may be buried at the crossroads, in a cemetery, or in the courtyard of a house. When one is initiated "on the point," the reference is to the condensation of the power of a particular spirit who has been diagnosed as the *mèt tet*.

The fourth and final level of initiation is the one that gives a person license to begin practicing as a healer. It is called *assògwe*, literally, "with the *asson*," the beaded rattle that gives priests and priestesses some measure of leverage in the spirit realm.

In Haitian Creole, the verb *kouche* (to lie down, to sleep, to make love, to give birth—less commonly, to die) is the general word used to describe initiation. Entering the initiation chamber is like dying. Friends and family members cry as they line up to kiss the initiates goodbye. Shortly after this genuinely emotional leave-taking, the initiates are blindfolded and led through a dizzying dance of spirals and turns before being taken into the small room where they will *kouche*. As in many other sorts of initiation around the world, to *kouche* is to be forced by ritual means to regress, to become a child again, to be fed and cared for as a child would be, only to be brought rapidly back to adulthood, a new kind of adulthood, again by ritual means. When the initiates leave the inner chamber after days of seclusion and ritualizing, they have their heads covered. Initiates must keep their heads covered for forty days. Like newborn babies with vulnerable soft spots, new initiates must protect the tops of their heads. The spirits within have been fed and are still changing and strengthening day by day. On an altar inside, the initiates have left their *pò tet* (head pots), residues of the internal externalized, the self objectified, the spirits concretized.

These *pò tet* generally remain on the altar of the priest or priestess who performed the initiation and who will be ever after the initiates' spiritual mother or father. Thus, through initiation rites, bonds among the living—as well as between the living and the spirits—are reinforced.

The Vodou Spirits

In the preceding discussion, I have been using the term "spirit" in a generic sense, as the Haitians often do, to refer to what are in fact three distinguishable groups: the *mò*, the dead; the *màwasa*, the divine twins; and the *mistè*, the mysteries, more often referred to as the *lwa*, or, using the term in a more specific sense, the *espri*, the spirits. Generally speaking, the dead and the divine twins are more central to rural than to urban Vodou. As the structure of the large extended families unravels, the sources from which people seek wisdom and assistance change. In the cities, possessions by specific powerful ancestors decline, while more energy is focused on possessions by the major Vodou *lwa*, most of whom are known and venerated throughout Haiti. In similar fashion, as children lose some importance for the work of the family, the divine children, the *màwasa*, also lose some ritual significance. However, neither the dead nor the *màwasa* disappear completely in the urban context.

The dead are still venerated in the cities. As was mentioned above, the *lwa* are inherited in urban families, where they will be remembered for some time as the *lwa* of a particular ancestor, for example, Marie's Ogou. Also, in the urban context family graves continue to be important, as do the annual celebrations for the dead that occur on and near All Souls' Day.

The *màwasa* also continue to have a role in urban Vodou. In addition to being routinely saluted in most large dancing and drumming ceremonies, the divine twins are given special attention in two contexts, both of which have to do with enhancing the luck of a particular group or a particular enterprise. The first instance has to do with making a *promès* (promise). This is done when resources do not permit the immediate fulfillment of an obligation to the spirits. In such a case a small *manje màwasa*, a meal for the divine twins, can be prepared. The dishes, favorites of children, will be fed to the actual children in the group. When they take obvious pleasure in the food this is taken as a sign that the spirits have agreed to accept the promise.

The second ritual in which the *màwasa* play a central role is the *manje pov* (feeding of the poor). This ritual is performed by families, both biological ones and those created around the urban Vodou temples. Ideally it is performed annually to ensure the good fortune of the group. Large quantities of all sorts of food are prepared. A small portion of this—a pot of soup, perhaps—along with coffee, soap, tobacco, and small change, is then sent to a gathering place for the poor. The steps of a church or the cemetery are likely places. These things are passed out to the poor along with an invitation to come to the temple or the home later in the day for a feast. Before any of those later assembled can eat from the overflowing pots prepared for the ceremonial meal, the children of the poor (a group doubly identified as the socially vulnerable) must first consume a separate *manje màwasa*.

Within the realm of the spirits, the *màwasa* play a role parallel to that of children in the social realm. They require more in terms of care and material goods than they can give back in the same media of exchange. However, because children are closely associated with the good fortune of a family as well as with its vulnerability (youngsters are said to be the most likely to "catch" destructive spirits sent against a family by its enemies), the exchange can be kept more or less balanced by the luck or blessing that children can uniquely bestow.

The *manje pov* reveals the connection that is made within Haitian Vodou between children and the poor. Both are socially vulnerable groups in need of care. Furthermore, the poor, like children, are understood to be sources of blessing. Almsgiving, particularly when on pilgrimage, is highly recommended in Vodou circles. The identical rituals that end both the *promès* and the *manje pov* reinforce the reading that helping children and the destitute brings good fortune. When the respective meals are finished, the guests—in one case the family children, in the other the poor, both children and adults—wash their hands in a basin containing water and basil leaves. The donor of the meal then stands in the center, and all guests wipe their hands on his or her clothing, face, arms, and legs.

By far the largest proportion of resources, time, and energy in the urban Vodou context is expended on service to the *lwa*. These *lwa* are both related to and different from their West African progenitors. The religious systems of the Fon and the Yoruba, both of which made central contributions to Haitian Vodou, have complex pantheons of spirits. These spirits have hegemony over a wide variety of life domains, including natural phenomena such as thunder, wind, rain, and smallpox, as well as cultural activities such as farming and hunting. When these rich spiritual systems were transported to the Caribbean, their considerable power to make sense of the world came to focus almost exclusively on the most problematic arena of life there, the social arena. For example, Shopona, the powerful Yoruba figure associated with smallpox, was completely forgotten. Others similarly associated with the powers of nature were lost unless their skills and proclivities translated readily into the social realm. In related fashion, many spirits were redefined in the New World setting. The Yoruba Ogun (the Fon Gu), a patron of metalsmithing, hunting, and warfare, came to be understood exclusively as a warrior in Haiti. This pervasive socialization of the divine occurred when West Africans were brought to the New World, and it happened again in new ways when their descendants were forced from rural homelands into the cities. Among the Gède (generalized spirits of the dead) recognized in Port-au-Prince are an automobile mechanic, a dentist, and a Protestant missionary. And Azaka, a *lwa* who is a peasant farmer, functions in his urban incarnations mainly as a voice reminding the dispersed of the importance of maintaining contacts with the extended family.

In the Haitian countryside (probably to a greater extent in former times than now) the various *lwa* are organized into several *nanchò* (nations). The names of these—for example, Kongo, Ibo, Wangol, Nago, Rada, Petro—almost all point to specific areas or groups in the African homeland. In the cities this complex of spirit nations has been synthesized into two major groupings, the Rada and the Petro. Within Vodou lore and practice these two groups are understood as fundamentally different, even oppositional. For example, mixing of the altars of the two pantheons is prohibited. Furthermore, even though both may be saluted in the course of a single evening, clearly articulated ritual transitions create buffer zones between the two groups.

The opposition between Rada and Petro can be best understood as a contrast between the quite different modes of relationship that each group represents. The Rada *lwa* are the "sweet" spirits. They are served with sweet foods and drink. The ambiance of their possession performances is intimate and warm. Even those Rada *lwa* who are awesome in their wisdom and power are treated with a respect that is transparent to the affection that underlies it. Rada spirits are *rasin* ("root") *lwa*. They are also said to be *frangine* (African). They are, in short, family, and the mode in which one serves them reflects this. While fidelity and caution are required in the service of the Rada *lwa*, these spirits are not overly strict in their dealings with the living. If a promised feast cannot be offered to them one year, they can be persuaded to wait until the next. The Petro *lwa* by contrast are characterized as "hot" spirits. Their possession performances often play at the border of violence and destructiveness. In like fashion, the unfaithful or careless devotee does not escape punishment. Why then would anyone serve the Petro *lwa?* Because they have access to realms of life that the Rada spirits do not. The power of the Rada *lwa* derives from their wisdom, including herbal knowledge. The power of the Petro *lwa* by contrast extends over, but is not limited to, the arenas of money and commerce. The Petro *lwa*, whose iconographic repertoire includes intricate and intense drum rhythms as well as police whistles, whips, and knives, are the spiritual incarnation of the plantation owners and their neocolonial equivalents—the mulatto elite who control the wealth of the country and the American and European businesspeople who profit from the labor of the poor. The opposition between Rada and Petro is thus aptly described as that between family members and foreigners, or insiders and outsiders. Not incidentally, the Petro *lwa* also chart a course for the person who would assert his or her individual needs over and against the demands of family. The two pantheons, Rada and Petro, thus offer different rewards and are in turn characterized by different modes of sociality. Relationships with spirits in both realms require reciprocity. However, exchanges with the Rada spirits take place in a warm familial atmosphere characterized by compassion, while those with the Petro *lwa* operate according to impersonal and inflexible rules and are thus pervaded with caution and anxiety.

The difficulty ethnographers have experienced in attempting to create a definitive list of the Vodou *lwa* is well known. The reason for this difficulty is rather simple: no such list is possible because the *lwa* are inherently mercurial. They are more accurately described as ways of being in the world, subject to endless transmutation through experience, than as beings per se. For example, the Haitians will say that there is one Ogou; they will also say that there are seven or twenty-one. In fact, there are probably many more than twenty-one that could be identified in the Port-au-Prince region alone. Each is an extension and elaboration of the central character of the warrior spirit Ogou. In his various manifestations Ogou plays across the full range of the constructive and destructive uses of power and aggression. For example, there is the politician Ogou Panama. There is the drunkard Ogou Yamson. There is Ogou Fèray the general, and Ogou Badagri the heroic soldier. Moreover, the individual personalities of the *lwa* are not exactly mercurial but similarly multifaceted. A particular *lwa* can exhibit power, dispense wisdom, and give solace and practical advice. But the same spirit can also—the particulars of his or her personality permitting—whine, pout, needle, harass, and become wantonly destructive. It is impossible, therefore, to group the Vodou spirits according to the moral categories of good and evil. Each spirit, Petro as well as Rada, has both constructive and destructive dimensions, and these change as the character of a *lwa* is applied to a particular life situation through the medium of possession performance. The *lwa* thus do not so much set examples for the living as they hold up mirrors that clarify certain aspects of the lives of those who serve them.

Treatment in the Vodou System

Vodou priests and priestesses treat a wide variety of *pwoblèm,* "problems." Clients come to them for help with love, work, and family problems as well as with sickness. The first determination that a Vodou healer must make is whether the problem "comes from God." If a problem is determined to have been sent by God, it is then seen as "natural" in the sense of that which is meant to be, that which is unavoidable.

When Catholicism blended with African religious traditions to create Vodou, the great West African sky gods, progenitors of human and divine beings alike, were absorbed into Bondyè (God). Bondyè (literally, the "good god") is the one and only god and is clearly distinguishable from the *lwa,* who are sometimes said to be his "angels." A popular Haitian proverb emphasizes the message that is contained in the name of god itself: "Bondyè bon" (God is good). As a result, if a problem, usually a physical illness in this case, is understood as coming from Bondyè, then it works to the greater good, even though this fact is unlikely to be apparent to the sufferer. No priest or priestess will interfere in such a case.

However, if a problem is determined to come from what some Haitians call "supernatural" causes, it is then thought to be appropriate for treatment within the Vodou system. It is important to remember that Haitians do not live in a two-story universe. God and the spirits are an intersecting dimension of life; they are not denizens of a separate realm. When they call a problem "supernatural," it means two things: the problem is not part of the natural order, meaning part of what is fated to be, and it is likely to have been caused by the spirits. Health problems that have a history of being resistant to scientific medical treatment often end up in the Vodou temple, where that very resistance is taken as a sign of the spirit-connected nature of the ailment. In fact, most problems are diagnosed as supernatural in origin or, if not specifically caused by the spirits, then at least falling within the province of their curative powers.

Once the preliminary determination is made that a particular problem is suitable for treatment, the *manbo* or *oungan* sets out to discover more about its nature and origins. Clients do not present themselves to Vodou healers with a detailed list of their symptoms. According to tradition, nothing more is required than a statement such as: "M'pa bon. M'pa genyen chans" ("I'm not well. I don't have any luck"). From this point, it is up to the priest or priestess to determine the nature of the problem, as well as its cause and cure. This is usually accomplished through divination.

The most popular form of divination used in Port-au-Prince is card-reading. However, gazing into a candle flame may be used or other more exotic techniques, such as pouring a small amount of alcohol into the top of a human skull and then reading the patterns made by the liquid moving along the cranial grooves—a very graphic appeal to the wisdom of the ancestors! For card divination, an ordinary deck is used with all cards below the seven removed. After lighting a candle and praying, the *manbo* or *oungan* offers the cards to the client for cutting. These are then laid out in four rows of eight in front of the healer. The whole process is repeated twice, once to determine the best description of the problem and once to track down its supernatural connections. After the first spread, the healer begins tapping the cards in patterns dictated by his or her own inner perceptions. Occasionally a question will be raised or a statement made. For example: "There is trouble in your house. I see fighting." The client is free to say yes or no without prejudice. Gradually, through a series of such statements and responses, the contours of the problem reveal themselves. It should be emphasized that while this is clearly not a miraculous procedure or even one requiring extrasensory perception, it nevertheless calls on the intuitive skills of the practitioner and represents an important step in the curing. When the problem is articulated through this gradual dialectical process, its definition may well surprise even the client. I once witnessed a session in which a mother brought her young daughter for help because the child would not eat, was losing her hair, and had run away from home. In the course of settling on the appropriate description of the problem, the *manbo* uncovered something that was unknown to the mother and unspoken before by the daughter: the girl's stepfather was sexually abusing her.

Once a full picture of the problem emerges, the healer then lays out the cards once more to determine its cause or origin: "I see the spirits love you a lot. Ezili especially. Did you promise you were going to do something for her and then not do it?" By this means a complete diagnosis is made.

Diagnoses point to disruptions in relationships. Often the relation in question is with the spirits themselves. Broken promises, lax or insufficient offerings, or refusal of the spiritual vocation the *lwa* have chosen for a person can all be reasons for trouble. Many *manbo* and *oungan* have dramatic stories to tell about their own efforts to resist the desire of the *lwa* that they take the *asson*, that is, undergo initiation to the priesthood. One woman was hospitalized three times and given last rites on two occasions for an intestinal disorder, the cause of which medical doctors could never determine. (Eventually she obeyed the *lwa*, and thereafter she reported that she experienced no further health problems.) Obligations incurred or promises broken by family members generations back can emerge as the cause of the contemporary individual's troubles.

However, as was seen in the case of the sexually abused child, it is not always the spirits who cause a problem. For example, the cards often reveal that someone is suffering because of the "jealousy" of other persons. Jealousy is understood to be such a strong emotion that the lives of its targets can be seriously disrupted. Within the Vodou system the object of jealousy rarely escapes at least part of the burden of blame. Such an attitude reflects a society in which it is expected that anyone who has much should give much. Thus, a wealthy person is almost by definition thought to be stingy, and a very lucky person is suspected of having done "work with the left hand." A less serious but related diagnosis is that someone is suffering from "eyes." This mildly unsettling condition comes from the fact that too many people are paying attention to that individual. It may be that there is gossip circulating. With both jealousy and eyes, as with several other diagnostic categories, the troubled relationships are among the living. In such situations the spirits are called on for help, but there is no sense in which they are seen as causing the problem.

Sorcery and Ethics

Disruptions in relations with the spirits cause serious problems, yet in many ways it is an even more serious situation if, in the course of a "treatment," it is discovered that a person's problems arise from the fact that another human being has done "work" against them. The range of magical actions that fall under the category of "work" is considerable. It may only be that a rejected lover has gone to the *manbo* or *oungan* for a love charm, or it may be something more serious, such as an act of sorcery performed by a vengeful neighbor.

For example, sorcery is frequently implicated when a diagnosis is made that a woman has "fallen into perdition." "Perdition" is a condition that befalls a pregnant woman in which the child in her womb is "held" or "tied" to prevent it from growing. When a woman who has missed one or more menstrual periods and assumes herself to be pregnant experiences a discharge of blood, she suspects that she may have "fallen into perdition." In all pregnancies it is believed that the menstrual blood that would ordinarily exit from the body each month is held in the womb where it serves as nourishment for the child. In a state of perdition the nourishing blood bypasses the fetus. The fetus, however, is not expelled but held inside the mother. Fetuses are believed to be able to stay in a state of arrested growth for years until something is done to "cut off" the perdition or "untie" the child. When that is accomplished the monthly blood flow stops, and the child begins to do its "work" within the womb. The infant born nine months later is the one who was conceived before the state of perdition began. Falling into perdition can be caused by several things. It can be caused if "cold" is allowed to enter the womb. It can be caused by restive *lwa* or ancestral spirits. However, work of the left hand, specifically sorcery, is the most frequent diagnosis. All children, but especially the unborn, are said to be susceptible to being "caught" by a work of sorcery directed against a family.[7]

There is an underlying belief in what might be called an economy of energy in Haitian attitudes toward sorcery or the work of the left hand. A rather flat-footed way of articulating the content of this belief would be to say: nothing comes for free. For example, there is a significant distinction made in the types of powers that a person can call on for help in this life. There are first of all *espri fami* (family spirits), and then there are *pwe achte* (literally, "points that have been purchased"). Most often residing in some tangible object such as a stone or bottle (the "point"), these spirits are either the souls of persons who died without family, ceremony, or burial, or they are the free-floating spirits of another, often malevolent, sort.

Serving family spirits entails obligations that may strain resources and energy; however, the demands of family spirits theoretically never escalate beyond reason. Within a given family the living and the spirits are interdependent in a way that makes both parties exercise restraint. Powers that have been purchased are another matter. While it is understood that they may be extremely effective, they have neither history nor loyalty to curb their rapacious appetites. Consequently, working with the left hand leads all too easily to an ascending spiral of obligations. Stories are frequently told of *manbo* and *oungan* who turned to sorcery in a desperate moment and then found it impossible to extricate themselves. First they lost members of their family; finally they lost their own lives. This belief that a person ultimately pays for what is gained through illegitimate means is one moral force within Vodou that curbs the wanton practice of sorcery.

Another moral force is the belief that only in extreme circumstances may one use sorcery to harm another, and only if one is absolutely just in doing so. For example, there was a *manbo* who lost her home through the deception of a woman friend who stole the title papers. The former friend actually went to court in an effort to claim the house for herself. The *manbo* performed a very simple act of magic (there is a widespread belief that the simplest ritual acts are the most powerful)[8] that involved dropping a "point" or charm into a latrine. As a result of this, three people either fell sick or died: the judge, the lawyer, and the erstwhile friend. When this incident was discussed within the family, someone invariably noted that the *manbo* could do this with no fear of reprisal from humans or spirits because she was so clearly in the right. The house was hers.

Yet another belief that acts to curb destructive uses of spiritual power centers on that part of Vodou associated with cemeteries. Although a version of this system operates within the cities, the pattern is clearest in the rural areas where cemeteries are still family property. The first male to be buried in a cemetery is known as the Baron Simityè, Baron of the Cemetery. When a wrong has been done to an individual or family by someone from outside that group, a simple ritual performed in the cemetery calls on Baron to send a *mò*, one of the souls of the dead, to avenge that wrong. The Baron's power can never be used, by definition, by one family member against another.

What complicates this discussion of morality and the uses of power within Vodou is the fact that it is not always possible to keep the categories clear and distinct. What is sorcery from one person's perspective is no more than what was required for an effective treatment from another's point of view. For example, love magic may heal a broken heart or soothe wounded pride, but it also necessarily involves the manipulation of the will of another. Cemeteries in Haiti are littered with the evidence of this common sort of "work." Small male and female rag dolls bound face to face and stood on their heads (inversion creates change) in a jar or drinking glass are evidence of a work designed to bring about a reunion. The same dolls bound back to back indicate that the dissolution of a troublesome relationship was the desired result. One bound with its face to the back of the other is said to be in a position to "eat" the other, that is, to take revenge. Such routine magic is within the repertoire of most Vodou healers and does not involve trafficking with suspect or "purchased" spirits.

Understandably, most priests and priestesses claim to eschew the work of the left hand. Equally understandably, rumors circulate that this one or that one "serves with both hands." It is not unlikely that most sorcery rumors can be attributed to individuals or groups in conflict wherein each party, knowing their own spirituality to be rooted in family and tradition, can only assume that the practices of their enemies are not so rooted.

Knowledge and Power

In the course of treating a troubled person, Vodou priests and priestesses call on a variety of different types of knowledge and power. The word *konesans* (knowledge) is used to refer to learned skills such as herbalism and divination as well as to what might be called intuitive powers. The different degrees of initiation are seen as increasing *konesans*. At least part of what is meant by this is sensitivity to a sense of foreboding. The attuned person, the one with *konesans*, knows when to cancel a trip or a business appointment. At a higher level of development it may be the gift of "seeing" what is wrong with people just by looking at them. (Although called seeing, one *manbo* described its physical manifestation as a prickling in the scalp.) Many of the most sought-after Vodou healers are said to have this gift.

In addition to their own developed talents, priests and priestesses also call on a range of higher authorities in the healing process. Possession allows the healer access to the awesome wisdom and power of the *lwa*, and in fact it is often one of the *lwa* who prescribes the specifics of a cure. Quite detailed information about what should be done to treat a particular case can also come in dreams. One *manbo* said that it is usually her dead mother (a powerful *manbo* herself) appearing to her in dreams who provides the solutions to her most difficult cases.

Dreams can also function in healing ways in the lives of ordinary devotees. Dreams can give warnings about bad things to come, thus providing the means of possibly avoiding sickness or anger, robbery or accident. Both the dead and the *lwa* routinely appear in dreams to give warnings and advice. The spirits sometimes appear in dreams in the same form as they are depicted on Vodou altars. Individual *lwa* have been conflated with particular Catholic saints, and the inexpensive and popular chromolithographs of the saints have thus become the most common images of the spirits. However, it seems that even more frequently the *lwa* appear in dreams in disguise. Each dreamer has his or her own code which must be applied to interpret the dream. Often it is a friend who has a name or personal qualities reminiscent of the *lwa*'s who comes to stand for that spirit in the dream world. Thus one *manbo* said: "Last night I dreamed about Gerard. [Saint Gerard is the Catholic saint conflated with Gède, the spirit of death.] Gerard asked me how my daughter was doing, if she was out of the hospital yet. That is when I got scared for my daughter. I was afraid she might really get sick because I know everytime I dream about Gerard, that's Papa Gède."

The care given by Vodou healers ranges from truly awesome displays of power to tender solace. I know of one *manbo* who brought her severely depressed female client into her home as part of the curing process. The woman had not spoken for nearly a year following the loss of a child. This mute condition, well known in Haiti and generally seen in young women, is considered especially difficult to treat. In the early stages of the treatment the *manbo* actually took the woman into her bed and held her until she slept. Yet treatments can also involve humiliation (e.g., being sent to the market to beg) and angry lectures from the spirits. In my experience, women healers routinely use the full range of care, from the solacing to the jarring, that is possible within the Vodou system. Male healers, by contrast, tend to remain authority figures throughout the healing process.

From a more general perspective, the jarring or confrontational aspects of Vodou healing are never separated from the overall context of familial care in which healing takes place. In fact, to make the distinction is to miss the coherence of the system. An image drawn from Haitian culture may make it easier to articulate this subtle point about the tone or ambience of caring within Vodou. Haiti is a child-centered culture. There are no events from which children are excluded. Yet the crying of infants and the misbehavior of older children are not tolerated. Crying babies are grabbed and rather roughly jostled into silence with unspoken messages that communicate at once the full attention of the caretaker and that person's unwillingness to tolerate the behavior. Older children can be given a harsh reproof at one moment and then a quick hug and kiss soon after. In a similar way traditional healers in Haiti can be possessed by an angry *lwa* without having that anger shape their personal relationship to the person seeking the cure.

The Creole verb *balanse* (to balance) has a special significance in Vodou and in healing within Vodou. When devotees take ritual objects off the altar they are instructed to *balanse*, to swing the objects from side to side. This is thought to awaken or enliven the objects and the spirits associated with them. The word can, however, be used in less constructive contexts. For example, when death touches a family it is said to "balance their house." The sense that balance is a dynamic condition is revealing, as is the notion that it comes out of opposition, whether that be the back-and-forth motion of the ritual *balanse* or the harsher clash of death against life. Within the Vodou view of things life is stirred up through opposition. This stirring and jarring, which can wound, is nevertheless healing when the clash of opposites is wisely orchestrated by the Vodou healer.

One example of a specific problem and cure will illustrate the confrontational dimension of Vodou healing. A young woman came to a *manbo* distraught, in fact nearly hysterical, because her husband had left her. In one moment the woman said she wanted her husband back; in the next she recounted a long history of his abuse. Finally, with a shrug of impatience, the *manbo* said harshly: "Pran tèt ou!" ("Get ahold of your head!"). Three ritual baths were prescribed to be administered, one each week for the next three weeks. The first bath was made from warm milk in which cinnamon sticks had been steeped. About four cups of the liquid were placed in a small enamel basin and the woman was instructed to remove her clothes. Because this was a good luck bath, the liquid was applied to the body from bottom to top, starting at the feet and stroking upward. (The reverse would operate in a bath designed to remove bad luck, a more serious condition.) The second bath was composed of various liquors and perfumes. It was applied in a similar fashion, as was the third and final glorious combination of champagne, roses, and perfume.[9] After each treatment the woman was instructed to leave the infusion on her skin without washing for three days. The first bath, she reported, made her smell of sour milk "like a baby." After it she took to her bed and cried for most of a week. She said that the second bath, in which alcohol was the dominant ingredient, burned her eyes and genitals. The second ended the tears, but she was flooded with anger. She sought out her former husband and screamed and yelled at him until the neighbors intervened. She reported nothing remarkable from the third bath beyond the fact that she no longer felt so unhappy. This sequence of baths took a woman's ambivalence about the man in her life and concretized it. The first and second baths shook loose contradictory emotions; they jarred her into powerful and direct experiences of sadness and anger. From the resulting dynamic "balance" came the possibility of the third bath, which moved her beyond the extreme moods of the first two to a less precarious emotional state, one in which she gradually was able to let go of the destructive relationship. These baths, like so many of the Vodou treatments, can also be seen as a ritual regression, a regression to infancy and then a movement back, or even as a ritual rebirth not entirely unlike that which is accomplished through the initiation ceremonies.

Conclusion

"Moun fèt pou mouri," people are born to die—the saying reveals the Haitian's sense that life is both short and painful. This verdict cannot change; it can only be accepted. Yet in the midst of the struggle that is life it is possible to enhance one's *chans* (luck) and minimize the *mizè* (suffering). This is accomplished in two ways: first, by respectful attention to the web of sustaining human relationships that defines family, and second, through conscientious service to the spirits who are after all members of one's own extended family, even—from one perspective, at least—parts of oneself. The spirits are served by the parent (fictive or actual) in the name of the family. In order to serve the family well in this role, the priest or priestess must have *konesans*: knowledge, intuition, insight into human and spiritual affairs. Such knowledge is most often rooted in the *oungan*'s or *manbo*'s own experience of suffering. To *kouche* (lie down, sleep, give birth, die, and, specifically, to be initiated) is to take the risk necessary to be healed oneself and through that process to enhance and focus one's power and knowledge in order to heal others. Once gained, *konesans* carries with it a moral obligation that it be used justly and respectfully. Thus, the *manbo* or *oungan* is one who knows how to *eshofe*, to raise the life energy in individuals and groups, human and divine. Power thus mobilized can then be concentrated in *pwe* (points) which are the concrete embodiments of relationships human and divine. Problems properly articulated in the concrete can be healed. One can pick up the *pwe* and *balanse*—turn the point upside down and bring about change that heals.

Notes

1. Claude Lévi-Strauss, chapter 1, "The Science of the Concrete," in *The Savage Mind* (Chicago: University of Chicago Press, 1966), 1–33.
2. As will be seen below, there is a sense in which the dead continue to exist; however, none of the living would consider this existence superior to his or her own. Thus immortality does not function as a reward for sacrifices made in the present life.

3. A partial qualification to this characterization exists in the large numbers of homosexual priests who have genuine power and prestige within Vodou. This is somewhat surprising given the extreme homophobia in Haitian culture. However, it is only a partial qualification because many of these priests are more accurately described as bisexuals. They often have traditional families.
4. See William Bascom, *The Yoruba of Southwestern Nigeria* (New York, 1969); also Melville Herskovits, *Dahomey: An Ancient West African Kingdom*, 2 vols. (Evanston, IL: Northwestern University Press, 1967).
5. Marriage to a Vodou spirit—a ritual complete with marriage license, an exchange of rings (wherein the spirit is represented by his or her *chwal*), a wedding cake, and, on occasion, champagne—is a ritual that does not demand that a person experience possession. It nevertheless involves a life-long commitment to the spirit. One day a week is dedicated to the spirit spouse. Special colors sacred to the *lwa* must be worn on that day, and the devotee must sleep alone so that the spirit may appear in dreams.
6. "Taking the *asson*" as a path to gaining status as a priest or priestess is a ritual performed mainly in the south of Haiti and in Port-au-Prince. In the northern part of the country such status is conferred by virtue of family position or reputation as a healer. The initiation rituals are costly for those who take the *asson*. It may be partly as a result of economic factors that individuals sometimes claim to have received priestly training in dreams, visions, or periods of time spent "under the water."
7. Gerald F. Murray, "Women in Perdition: Ritual Fertility Control in Haiti," in *Culture, Natality and Family Planning*, eds. John F. Marshall and Steven Polgar, 59–78 (Chapel Hill, NC: University of North Carolina Press, 1976).

   Murray points out that the socially useful part of this explanatory scheme is that, in providing the possibility of a pregnancy much longer than nine months, a woman can claim the father of her child to be almost anyone with whom she has ever had sexual relations. This in turn allows her to choose among fathers the one who is most likely to be able to give meaningful support. Given the current social instability all over Haiti, finding men with the means and temperament to be responsible fathers is one of the major problems faced by women.
8. See Serge Larose, "The Meaning of Africa in Haitian Vodu," in *Symbols and Sentiments: Cross-Cultural Studies in Symbolism*, ed. Ioan Lewis, 85–116 (New York: Columbia University Press, 1977).
9. The ingredients for Vodou treatments are paid for by the client. Fees for the healer beyond the cost of materials are understood to be gifts, and theoretically it is up to the client to decide how much he or she will offer. In practice, however, the range of what is appropriate is usually well known to clients without their asking. It is worth noting that many of the most sought-after healers are not prosperous persons. They adhere strictly to the tradition that healing powers are not to be used for inordinate profit.

Chapter 2

Vodou in Haiti: Way of Life and Mode of Survival

Claudine Michel

Si se pate bon Ginen sa-a,
nou tout ta peri deja . . .
If it were not for the Guinea lwa,
we would all have perished already
Ayibobo!!!

—*Fleurant, Dancing Spirits, p. 155*

This chapter originates in the movement of rediscovery and rehabilitation of religions and modes of spirituality of indigenous peoples with a long history of subjugation and whose beliefs have been dismissed continually as primitive if not downright evil. Animism, fetishism, paganism, heathenism, and black magic are some of the terms that have been used improperly in the West to describe the Haitian religion, which is presented in the foreign press and the media as a religion of blood and sacrifice, as a religion of sexual orgies and malevolence, thus resulting in the widely shared perception that the practice of Vodou equals sorcery and witchcraft.[1] This work is revisionist in that it recasts the values and principles inherent in the Vodou religion, in particular its humanism and sense of communality, and emphasizes the complexities behind the way these values are transmitted from one generation to the next in Vodou communities. It also shows how this New World religion combines and recombines the Africanisms from which it originates with its American and Creole realities.[2]

Vodou, presented in the West in opposition to *true religion,* does not, in effect contrast with *Western religions* as much as the media would like us to believe. The contrast is with the Christian churches and the established religions of the West, their dogma, phenomenology, hierarchy, and monumental architecture. What is ironic is that Haitians who are Vodou adepts perceive themselves as good Christians and see no conflict between practicing Catholicism and being a member of a Vodou family and community. These are two beliefs systems, two modes of functioning, two worldviews, which are separate yet interconnected by a common longing for the divine and a never-ending search for a better life for followers and devotees.

A serious analysis of the Vodou religion as practiced in Haitian society and abroad reveals that Vodou is a broadly encompassing worldview, a comprehensive system that shapes the human experience of its adepts in their search for higher grounds and purpose in life. Despite an apparent absence of a formal church and clergy, of written dogma and other such publication or instructional material, the Vodou religion is omnipresent, pervasive, strong, and performs key functions in all aspects of Haiti's social and political life. As such, the ancestral religion represents a key element of Haitian consciousness and provides moral coherence through common cosmological understandings.

Many of the monotheistic religions—Christianity, Islam, Judaism—are prescriptive and accompanied by a book of law. From sacred texts, followers derive a corpus of doctrines and extract general laws as measures of moral quality. Things are more complex in Vodou, which offers few absolutes and generalities, only trends and thematic modalities of lived ethical life. However, the religious tenets of Vodou, though not written in a sacred book, are not at all different from what is prescribed by the Christian faith, Judaism, Shinto, Hinduism, or Buddhism, among other religions.[3] Like members of other persuasions, those who participate in Vodou life and rituals believe in creating harmony, in keeping a balance, in cultivating virtues such as justice, beneficence, benevolence, patience, forgiveness, and cooperation, in respecting elders, and instilling desirable values in children.[4]

Vodou is not a system imposed from above; it is a democratic and functional religion, embedded in the vicissitudes of its followers' daily existence, and in their struggle for survival. In Vodou, as in African and other non-Western traditions, there is complete unity of religion and life. Vodou is central to the Haitian experience and as such cannot be abstracted from the day-to-day life of its followers. It is based on a conception of reality that includes life's goals, the forces that determine the fate of living things, proper social organization, balanced interpersonal relations, and practices that promote the welfare of the community of believers. Its devotees ask of it what people have always asked of religions: a basis for daily living, remedy for ills, help in times of hardships, satisfaction of needs, and hope. McCarthy Brown explains in her book *Mama Lola* how Vodou is a "repository for wisdom accumulated by a people who have lived through slavery, hunger, disease, repression, corruption, and violence—all in excess."[5] She further comments about Vodou's functional attributes: "Vodou is the system that they [Haitians] have devised to deal with the suffering that is life, a system whose purpose is to minimize pain, avoid disaster, cushion loss, and strengthen survivors and survival instincts."[6]

Above all, the Haitian religion is a practical faith whose primary concern is the well-being of individuals and the welfare of the group. People turn to the spirits and their ancestors to secure a better life for themselves and their community—that is, to assure the survival of the self and of the group in *this* lifetime. Unlike other religions,

Vodou does not have a concept of Eden or Heaven. Therefore, survival in this lifetime and *healing* for immediate well-being become an ongoing process that engages Vodou adepts throughout their life and turns into the primary goal of their existence.

Vodou as a Religious/Philosophical Persuasion

Writing about *Vodou*[7] and morality or *Vodou* as a form of humanism may appear paradoxical when *Voodoo*[8] suggests distorted images of superstitions, sorcery, blood sacrifices and sexual orgies. Scholars such as Hurbon, Laguerre, Bellegarde-Smith, Desmangles, Montilus, Fleurant, and others such as Thompson, Deren, Courlander, or McCarthy Brown, for example, have attempted to penetrate Vodou's true meaning and recast the real significance of the religion. Using new epistemological foundations and methodological approaches, their work aims at reconstructing and rehabilitating the essence of the Vodou religion, which in turn sheds light on those core traditional values sustaining Haitian social and political institutions.

Vodou is the harmonious fusion of beliefs and rituals of West African origin, which, having incorporated some Catholic practices, has come to be the religion of the greater part of the peasants and the urban proletariat of the Republic of Haiti. These are joined by members of the Haitian elite and some foreign nationals. Courlander writes, "In short, it is a true religion which attempts to tie the unknown to the known and establish order where there might otherwise be chaos. For those who believe in Vodoun, no event or episode is a thing in itself. In birth and death, good fortunes and bad, the Lwa[9] are somehow involved."[10] Bellegarde-Smith's encompassing definition of Vodou[11] sheds light on the true meaning of the religion. He explains,

> Vodun is a coherent and comprehensive system and worldview in which every person and everything is sacred and must be treated accordingly. In Vodun, everything in the world—be it plant, animal, or mineral—shares basically similar chemical, physical, and/or genetic properties. This unity of all things translates into an overarching belief in the sanctity of life, not so much for the *thing* as for the *spirit* of the thing. The cosmological unity in Vodun further translates into a vaunted African humanism in which social institutions are elaborated and in which the living, the dead, and the unborn play equally significant roles in an unbroken historical chain. Thus, all action, speech, and behavior achieve paramount significance for the individual and the community of which the individual is part.

Davis offers yet another useful definition, which again embodies the complex interconnections between the scared and the mortal while also capturing the day-to-day, functional aspects of the religion that regulate chaotic actions as well as moral and ethical conduct in Haitian society.[12] He views Vodou as

> a complex mystical world view, a system of beliefs concerning the relationship between man, nature, and the supernatural forces of the universe. Vodoun cannot be abstracted from the day to day life of the believers. In Haiti, as in Africa, there is no separation

> between the sacred and the secular, between the holy and the profane, between material and the spiritual. Every dance, every song, every action is but a particle of the whole . . . Vodoun not only embodies a set of spiritual concepts, it prescribes a way of life, a philosophy, and a code of ethics that regulate social behavior.[13]

The followers of the ancestral cult refer to their religious beliefs and practices by the phrase *sèvi lwa yo*—which can be best translated as "serving the spirits."[14] An adept of Vodou simply says, "I serve the spirits," which in itself is a revealing statement about the nature of the religion, the importance of withdrawing the *self* and serving *others*, and about the spiritual connections existing between living human beings, their ancestors, and their Gods. Also, an understanding and knowledge of African religions and philosophy allows one to read even more into such a phrase: it clearly connects this religion of the New World to the African Ethos and Worldview. Harold Courlander's compelling commentary further exemplifies Vodou's African connection, resilience, and pervasiveness:

> Vodou permeates the land, and, in a sense, it springs from the land. It is not a system imposed from above, but one which pushes out from below. It is a thing of the family, a rich and complex inheritance from a man's own ancestors. It is not the priests of Vodoun who control and direct its course. They, like the lowest peasant simply move about within it and make use of its resources. Vodou is strong and it cannot die easily. . . . You cannot destroy something with such deep genuine roots. You may warp it, twist it, make it crawl along the ground instead of growing upright, but you cannot kill it . . . [especially] in light of the inner history of the race.[15]

For Haitians, religion is life. Vodou permeates all aspects of their existence from the highest forms of interactions with the divine to the most mundane and profane matters. Vodouists bring their African/Creole spirits, their *Lwa* into all affairs and consistently seek their influence. As Bellegarde-Smith explains, the *Lwa* "represent the cosmic forces that are integral to the Haitian experience and yet transcend it."[16] On a daily basis, the religious life of those who *serve the spirits* (*sèvi Lwa yo*) revolves around a form of collective self-consciousness guided by their *Lwa* and the following African-derived principles: holistic conception of life, human-centered orientation, centrality of the community, honor and respect for elders, beneficence, forbearance, forgiveness, and sense of justice, all manifestations of a new Creolized African humanism. Robert Farris Thompson was indeed correct when he wrote that *"vodun* was Africa *reblended."*[17]

Humanism in Haitian Vodou

All knowledge presupposes a fundamental holism grounded in the idea of oneness and unity of all forces of nature, in the idea of interdependence and interconnectedness of these forces, and in the premise of supremacy of totality over individuality. The universe is a seamless cosmos where every force of nature has meaning and a connection with other entities. Creating dissonance in nature's polyrhythms, disturbing the harmonious flow of things, and bringing about division in the community are all acts that represent moral transgression in the Vodou world. Due to the web of interconnectedness, a person's moral violations distract, disturb, and perturb the outer world that ought to seek restoration of its harmonious state and rhythm. Morality for those who *serve the spirits* is a constant effort to maintain social cohesion, harmony, and balance. What is "right" in the Vodou world is not a function of abstract reasoning, but is relative to what will achieve *unity*.

Human beings occupy a central cosmic position in Haitian spirituality that is anthropocentric as well as humanist. Though people are not the "rulers" of creation, they are at the center of it, which explains why Vodou's paramount goal is to improve living conditions on earth. The veneration of deities is purposeful and serves humans through blessings extending from good health and well-being to marriage and love, to work and finances, to rain and plentiful harvest, and to children who grow up to be respectful of people and traditions. Human connection is the assumption in the Haitian worldview: there is suppression of unique life history in favor of a collective *personhood* from which energy is derived. Individuals become real only through interactions, and "it is precisely in responsive and responsible relation to others that one has the clearest and most steady sense of self."[18]

The paradigm *"we are, therefore I am"*[19] expresses well the "societary" essence of the Haitian community. It shows the ties that link members of the same community as part of the human web. We are not only accountable for our individual acts, but we are also responsible for people around us in that each of our actions and deeds influences the balance of the outer world. There are major communal expectations in the Vodou world and the extended family is the first unit where this communal sense manifests itself. It is a highly valued institution that prepares the individual for integration into the larger community. Ignoring family responsibilities, jeopardizing communal interests, and neglecting the *Lwa* are serious moral offenses that trigger the disapproval of the group and may endanger the care and protection of the spirits. Morality in the absolute sense can never be placed above the welfare of the collectivity.

Customs and implicit rules regulate obligations and responsibilities toward the collectivity and the entities that compose it. Allegiance, love, prayers, faithfulness, respect, material support, as determined by seniority, status, and financial means, are due to all members of the group under all circumstances. However, extreme respect and honor are particularly accorded to the elders, the bearers of knowledge and wisdom, the repository of experience. Respecting them, honoring them, and serving them is inherent in the very essence of the religion. To be virtuous is not only to provide for the elderly, to give them care, food, money, clothing, but also to show them love, respect, deference, and appreciation. Failure to give absolute respect to elders may attract imbalance in the life of the person who has departed from the traditional values of the society, bring about "bad luck" for her or his family and is morally destructive for the entire community.

Though morality in the Vodou world is fluid, dynamic, and always contextual as life itself, many other virtues are also valued for their potential to help develop knowledge and wisdom among adepts of the Haitian religion and to support one another. Among these traits are beneficence, forbearance, forgiveness, and a sense of justice, described here using the model of "virtue theory" elaborated by Peter Paris in his book *The Spirituality of African Peoples.*[20] He states, "Synonyms for beneficence are many. They include hospitality, generosity, liberality, benevolence, magnanimity, love. The beneficent person is a person of good will, one who joyfully extends hospitality to all alike."[21] About forbearance, Paris writes, "Because it helps people to survive dehumanizing conditions, forbearance can be one of the most effective means for preserving and promoting the goal of the community . . . Activities that serve the pragmatic goal of survival are often the habits necessary for developing the virtue of forbearance."[22] He comments in regard to forgiveness, "African peoples seem to exhibit such little racial hatred in return for the misery that they were forced to endure. Even during the most intense periods of resistance to slavery, segregation, colonialism, [or] apartheid, they have rarely been consumed by the spirit of hatred."[23] On the final trait, justice, Paris remarks that "it is the supreme virtue because it is the sum of all virtues . . . As with all virtues, African understandings of justice are regulated by the demands of their ultimate goal, namely, the preservation and promotion of community."[24]

The first section of this chapter shed light on the Vodou worldview and the moral principles embedded in the religion. In a work studying the persistence of African religions in the Americas, Barrett comments, "The slave master was able to claim the body of the slave, but the worldview of the African was nurtured in his soul and this soul was impregnable."[25] He explains how religion was the motivating and unifying force for the many different ethnic groups who found themselves in the New World[26] and acknowledges Vodou's "noble history" as a catalyst for the revolutionary accomplishments for which Haiti is known.[27] It is well documented that through their traditional religion and its worldview, Haitians survived oppression, found modes of expressions, and re-created a modified African society and ethos, which have to some extent Africanized the American continent. In what follows, the manners in which these values are communicated will be discussed.

Teaching through Words and Images, Action and Deeds

In Western societies, people write and read books to learn about past experiences and remember their ancestors. Those who *serve the spirits* are part of a world that apprehend historical, social, and religious experiences in a holistic manner, which then takes form in a lively and meaningful fashion, via "oral performance" through family memories, tales, metaphoric images, proverbs, songs, prayers, and various other spiritual and artistic expressions that represent an extremely rich repertory of practical wisdom. Liveliness, relevance, and functionality take precedence over truthfulness, clarity, and objectivity as paramount virtues.

Especially because 80 percent of the population cannot read, Haiti's rich *oral/social performance* tradition remains strong and ultimately necessary as a conveyor of the Haitian worldview and values. The indigenous tales, stories, proverbs, sayings, songs, and prayers acquire livelihood, as it is the case in all African traditions, because they have a purpose and a meaning. What counts is not the songs, tales, or words themselves, but it is their significance and relevancy to people's lives and immediate well-being.

However, despite the apparent importance of words and the oral medium for value transmission in Haitian society, the spoken language has very little meaning and practically no influence in itself; images, metaphors, contradictions, irony, humor, and tone are what give life and significance to the message behind the words. The "exercise of the art," that is, the *oral performance*—how things are said, by whom, to whom, under which circumstances, for what purpose—"actualizes" the words and gives them life. The following passage of Karen McCarthy Brown further exemplifies this notion of "words in action":

> Vodou is not a religion of words. Its wisdom does not lend itself to doctrinal formulation in which the truths remain the same no matter who speaks them or when or where they are spoken. The closest things there is in Vodou to theological discourse resides in the songs, songs which have a cryptic tricksterism to them that makes for double-, triple- and quadruple entendre. Not infrequently the various meanings are conflicting and contradictory. In the Vodou ceremonies in which these songs are sung, it is the energy of the drumming that creates a dynamic balance among the paradoxes contained in the words.[28]

In Vodou, like in other religious and secular popular manifestations found in other areas of the world, the concept of "art for art's sake" does not exist. Each word, each object has a function—often a liberating function, one that channels energy and empowers people against oppressive systems.

In the same fashion that *words* acquire meaning through action, *deeds* accompanied by words, songs, and music become more inclusive, more encompassing, and turn into a compelling creative production that exerts a strong educational influence. Typically, instructional strategies in Vodou involve the use of both words "acted out" and cultural elements and values that find expression in spoken language, body language, rituals, dramatizing, singing, and music. A person gives life to what is being acted upon and cosmic energy is derived through usage and practice. Thus, Vodou adepts are taught that a proverb, a song, a musical rhythm take life only when said, recited, or cadenced in the movement of the dance. Gerdès Fleurant may well have summarized this whole notion when he writes that "Vodoun . . . is a danced religion."[29]

As a result, a rich tradition of teaching and learning has developed in Haiti, encompassing uses of symbols, rituals, ceremonies, proverbs, wise sayings, memorizing, apprenticeship, storytelling, observation, practicing, singing, dramatizing, and sometimes writing.[30] All these activities are both acts of learning and instructional strategies since, as Maya Deren observed, Haitians have neither the time, the energy, nor the means for inconsequential activity. She remarks that in Haiti "[religion] must do more than give moral sustenance; . . . must do more than provide a reason for living, it must provide the means for living. It must serve the organism as well as the psyche. It must serve as practical methodology, not as individual hope. In consequence, the Haitian thinks of his religion in working terms."[31] As

Leslie Desmangles stated,

> Vodouisants cannot afford the self-surrender of mysticism, nor can they permit themselves the luxury of an idealism that seeks to mask the miseries and frustrations of their existence. Their needs are too immediate for that. Their religion must satisfy actual needs rather than merely invite them to high-flown intellectual exercises of theology.[32]

Philosophizing, discoursing, or praying with no intended purpose are not strategies typically used in Haiti's ancestral religion. The various forms of ritualizing and rituals all have a function. Vodou *services* represent an opportunity for the deities to share in the life of the Vodou family and are always *collective healing* ceremonies. Fleurant relates how the Vodou drummer TiBe explains *moun chofe*, that is, people get heated up, "the community warms up and becomes whole." Fleurant continues to describe this moment of liberation for souls and bodies—this true moment of freedom through communion of the devotees with their *Lwa Ginen*. He writes,

> Thus, the dialogue of drummers, dancers, and singers, through the kase and vire,[33] permits the participants to meet the spirits in possession, a state that liberates the soul and brings the community together in an eternal dance of freedom. And the people, remembering old, old things, sing and dance in unison: "Si se pate bon Ginen sa-a, nou tout ta peri deja" or, "if it were not for the Guinea lwa, we would all have perished already." *Ayibobo*!!![34]

The living enter in communication with the spirits and their ancestors to get advice on matters of importance for the members of the group and to seek assistance in all practical matters of life. In return for promised luck and protection for the family and its land, small tangible gifts of food or other sacrifices are offered to the *Lwa* in support and appreciation. These offerings strengthen the relationship between the living and the dead and allow the spirits to exert their cosmic power partaking of the very realm of human existence. In this respect, the down-to-earth names given to some of the ceremonies are quite revealing: "*manje les anges* [ritual feeding of the spirits/the *Lwa*]," "*manje marasa* [ritual feeding of the spirit of the twins]," "*manje pov* [ritual feeding of the poor]."

In Vodou, a person is meant to derive energy from interactions with others, therefore, all encounters with nature, with fellow humans, and with the spirits create opportunities for understanding, for growth, and for healing. Considering that everything has a soul, from the smallest grain of sand to the Cosmic Being, the dynamic force infused in the teaching and the learning of all tasks and ideas, material and spiritual, becomes a learning experience in itself. In Vodou, every dance, every song is a prayer; every word and every act becomes a lesson. In other words, one learns *everywhere* and *at all times*. More than actual "moral" principles, what is communicated inside and outside Vodou families is a worldview,[35] a particular sensitivity and perspective on life—the *Vodou ethos*—and various modes of survival to face the numerous oppressive situations that have become a way of life for Haitians.

This chapter would have fulfilled two primary functions by aiming at the following (1) providing knowledge about the nature of the Haitian religion, which is a religion *à part entière* and about how Vodou is, in effect, a means of survival for the Haitian people; and (2) emphasizing the need to understand both the Africanisms from which Haitian Vodou derives and its new Creole realities. The Vodou religion is in constant dynamic evolution, unremittingly creating and re-creating an internal modus operandi, blending African/Creole/Haitian elements and responding to the ever-evolving needs of its devotees. It is important to understand, as Jacob Olupona wrote about African traditional religions, that Vodou is not "just a house of cards that collapsed at the instance of change, but that it has the potential to adapt to change on its own, in response to changes taking place around it."[36]

Notes

1. The similarities between the following two definitions show that misrepresentation and misinformation about the Haitian religion have not changed in thirty years. "A body of primitive rites and practices, based on a belief in sorcery and the power of charms fetishes, etc., found among natives of the West Indies and in the southern United States, and ultimately of African origin," (*Webster's New World Dictionary*, 1964). "A primitive religion of West African origin, found among Haitian and West Indian Negroes and the Negroes of the southern United States, characterized by belief in sorcery and the use of charms, fetishes, witchcraft" (*New Illustrated Webster's Dictionary of the English Language*, 1993). Books such as *The Magic Island* (New York: Harcourt Brace, 1929) and others have been reprinted and continue to shape the general public perception about Vodou. Also, let's note the deleterious influence of a plethora of films made for American consumption such as *Angel Heart* or the Hollywood version of *The Serpent and the Rainbow*.
2. Charles H. Long, personal communication, Santa Barbara, CA, August 1996.
3. See Huston Smith, *World's Religions* (New York: Harper San Francisco, 1994); David Chidester, *Patterns of Action: Religion and Ethics in a Comparative Perspective* (California: Wadsworth, 1987); R. Murray Thomas, ed., *Oriental Theories of Human Development: Scriptural and Popular Beliefs from Hinduism, Buddhism, Confucianism, Shinto and Islam* (New York: Peter Lang, 1988).
4. See Claudine Michel, "Tapping the Wisdom of the Ancestors: An Attempt to Recast Vodou and Morality through the *Voice* of Mama Lola and Karen McCarthy Brown," Research Report No. 27, William Monroe Trotter Institute, University of Massachusetts, Boston, 1996; "Women's Moral and Spiritual Leadership in Haitian Vodou: The Voice of Mama Lola and Karen McCarthy Brown," *Journal of Feminist Studies in Religion* 17, no. 2 (Fall 2001): 61–87.
5. Karen McCarthy Brown, *Mama Lola. A Vodou Priestess in Brooklyn* (Berkeley: University of California Press, 1991), 98.
6. Ibid., 10.
7. *Vodou* is still the most commonly used spelling by those who write on the Haitian religion though, increasingly, the more correct spelling *Vodun* is being introduced in scholarly publications. The word is believed to derive from two West African words, Vo and du, which mean, according to some, introspection into the unknown, and to others, a form of reverence toward the supreme being. Some spell it *Vodoun* instead of Vodun. *Vaudou* is still used but mostly in France as Haitians tend to prefer Vodou.
8. In the United States, the word *Voodoo* is used in a casual and derogatory manner to indicate on the one hand anything from the deceptive to the downright evil (as discussed earlier) and on the other hand anything magical or miraculous (this use of the term is recent). Examples include Voodoo economics, Voodoo politics, Voodoo Mac.

9. *Lwa* or *Loa*(s) are the Vodou spirits of the Haitian pantheon. They serve as intermediaries between the ultimate God, Bondye, and the humans. Each *Lwa* is an archetype of a moral principle that he or she represents. Among the *Lwa*/spirits most frequently "served" and invoked in ceremonies are the following: Dambala, supreme, oldest, most respected, represented by a snake; Aida Wedo, his wife; Legba, the spirit of the crossroads who must be invoked to "open the gate" for the other *Lwa*; Ogou who does not tolerate injustice, but who can be mean at times; Erzili, representing sexuality, lesbianism, motherhood; Azaka, the peasant, the worker, the one who controls money; Baron Samedi and Gran Brigit, guardians of the cemeteries; Gede, the spirit of death and sexuality.
10. Harold Courlander, *The Drum and the Hoe* (Berkeley: University of California Press, 1973), 12.
11. Patrick Bellegarde-Smith, *The Breached Citadel* (Boulder, CO: Westview Press, 1990), 13.
12. Though Wade Davis's work offers at times valuable insights into the religion, his work is considered controversial in a number of ways. Some wonder why he allowed Hollywood to make the film *The Serpent and the Rainbow*, which is such a poor rendition of his book *The Serpent and the Rainbow* (New York: Warner Books, 1985). Also, *Passage of Darkness: The Ethnobiology of a Haitian Zombie*, raises questions—at least for some—about methods of data collection and the revelation of such shielded material to the general public.
13. Davis, *Serpent and the Rainbow*, 72–73.
14. Haitians do not say that they are Vodouists or that they follow or practice Vodou, they use the Creole phrase "*sèvi lwa*," which means to "serve the spirits."
15. Courlander, *Drum and the Hoe*, 7.
16. Bellegarde-Smith, *Breached Citadel*, 22.
17. Robert Farris Thompson, *Flash of the Spirit, African & Afro-American Art & Philosophy* (New York: Vintage Books, 1983), 164.
18. Karen McCarthy Brown, "Alourdes: A Case Study of Moral Leadership in Haitian Vodou," in *Saints and Virtues*, ed. J. S. Hawley, 162 (Berkeley: University of California Press, 1987).
19. John Mbiti, *African Religions and Philosophy* (Nairobi, Kenya: East African Educational Publishers, 1969; repr., 1992); *Introduction to African Religion*, 2nd ed. (Nairobi, Kenya: East African Educational Publishers, 1992).
20. Peter J. Paris, *The Spirituality of African Peoples: The Search for a Common Moral Discourse* (Minneapolis, MN: Fortress Press, 1995). Specifically, his model of "virtue theory" can be found on 130–156.
21. Ibid., 138.
22. Ibid., 141.
23. Ibid., 148.
24. Ibid., 152–153.
25. L. Barrett, "African Religions in The Americas: The 'Islands in Between'," in *African Religions. A Symposium*, ed. N. S. Booth, 184 (New York: Nok Publishers, 1977).
26. It is reported that in Haiti alone over a hundred different ethnic groups found themselves on the island and were able to unite only through the African re-created religion, Vodou, which somewhat compensated for their losses (family, clan, land) and brought them closer to Guinea—a word that in Haiti is used until today to refer to Africa in general, to a paradisical vision of Africa that one finds, for instance, in many Haitian paintings, and to what they see as the positive side of Vodou, the work done with the "right hand" (as opposed to the work done with the "left hand," the more destructive aspect of Vodou). Vodouists will often say "*mwen sèvi Ginen*" or "*Ginen yo*," which roughly translates as I serve the spirits, the *Lwa* from Guinea, Africa's good deities.

27. Barrett, "African Religions in The Americas," 198.
28. Karen McCarthy Brown, "Plenty Confidence in Myself: The Initiation of a White Woman Scholar into Haitian Vodou," *Journal of Feminist Studies in Religion*, 3, no. 1 (Spring 1987): 74.
29. Gerdès Fleurant, *Dancing Spirits: Rhythms and Rituals of Haitian Vodun, the Rada Rite* (Westport, CT: Greenwood Press, 1996), 10.
30. See Elleni Telda, "Indigenous African Education as a Means for Understanding the Fullness of Life: Amara Traditional Education," *Journal of Black Studies* 23, no. 1 (Fall 1992): 7.
31. Maya Deren, quoted in Leslie Desmangles, *The Faces of the God. Vodou and Catholicism in Haiti* (Chapel Hill, NC: University of North Carolina Press, 1992), 5.
32. Ibid., 5.
33. *Kase* in *Yanvalou*—a most important Vodou dance—means to bend one's knees, to bend one's back as opposed to dancing upright. *Vire* is a Creole term, which means to turn around.
34. Fleurant, *Dancing Spirits*, 155.
35. Claudine Michel, *Aspects Moraux et Educatifs du Vodou Haïtien* (Port-au-Prince, Haiti: Le Natal, 1995).
36. Jacob Olupona, "Major Issues in the Study of African Traditional Religion," in *African Traditional Religions in Contemporary Society*, ed. Jacob Olupona, 32 (St. Paul, MN: Paragon House, 1991), 32.

Chapter 3

African Interpretations of the Christian Cross in Vodou

Leslie Gerald Desmangles

The contact between Dahomean religions and European Catholicism in Haiti during the colonial period (1492–1804) resulted in a system of correspondences between these two religions. Such correspondences can be seen in Vodun not only in the reinterpretation of many Catholic saints but also in the reinterpretation of the Christian cross. Contrary to the opinions of many scholars who claim that the use of the Christian cross in Vodun ceremonies is Christian in character, this article points out that Vodunists do not interpret the symbol of the cross in the context of Christian theology, but in the context of Dahomean mythology.

Scholars who have written about Vodou, the folk-religion of Haiti, have called it a syncretism, that is, an amalgam of various religious traditions (Price-Mars 1928; Jahn 1961; Herskovits 1971). These traditions permeate the folk-beliefs and ritual dances of the peasant masses of Haiti. The religious amalgam originated with the arrival in the New World of slaves from West Africa as early as 1510.

In reading Moreau de Saint-Méry (Saint Méry 1797, vol. I:45–58), an eyewitness of the slave trade, one finds an impressive list of the slaves and the descriptions of the character of the tribes represented on Haitian soil. One might deduce from this list that all of Africa contributed to the formation of Vodou. However, from other writings of the slave period, it is quite clear that the region of the Gold Coast, particularly that of Dahomey, was the main provider of slaves. It is in this connection that Jaheinz Jahn, paraphrasing from a portion of *The Reports for Consideration of all Matters Relating to Trade and Foreign Plantations* writes:

> The reason why it was the religious conceptions of Dahomey in particular that came to prevail in Haiti is apparent from a London report which tells us that ten to twelve thousand slaves were exported yearly from the kingdom of Dahomey. The English exported

> only seven to eight hundred of these, the Portuguese about three thousand and the French the remainder, in other words more than six to eight thousand a year, who were shipped to the French Antilles, above all to Saint-Domingue, as the principal French colony of Haiti was then called. (Jahn 1961, 29–30)

Jahn's thesis of the Dahomean influence in Haiti is reinforced by the following facts. The first is linguistic in nature. Many of the words in the religious vocabulary of Vodou are Dahomean and have remained uncontaminated by plantation life. To cite only a few, *Rada*, a word which designates one of the Vodou pantheons, is a contraction of *Arada* (or *Alladah*), the name of a Dahomean city. This pantheon includes such *loas* (Vodou deities) as Legba, Damballah Whydah, Aida Whydah, Agwê, Sogbo and Aizan, all of whom are deities of the sky pantheon in Dahomey. The word *Dahomey* itself figures in the name of such Rada *loas* of fertility as Ezuli Fréda Dahomey, where *Fréda* is probably a corruption of the word Whydah, an important Dahomean seaport.

The word *Vodou* in both Dahomey and Haiti means the "company" or the "family" of the gods. Furthermore, the source of the root *hun-*, which figures in the Vodou terms *houngan* (Vodou Priest), *hounsi* (initiate), and *hounfort* (temple) is clear, for in Dahomey it is synonymous with *vodou*. The origin of *houngan* and *hounsi* are perhaps less clear than *hounfort*, for in Dahomey *-gan* in houngan means "priest" or "chief of spirits" and *-si* in *hounsi* means "wife," employed in the sense of the "wife of the deity" or again a devotee. Thus, Haitian Vodou as a religion concerns itself with establishing communication between the *loas* and their devotees.

The second fact which confirms the influence of Dahomean religion on Vodou is a theological one. In Haiti, as in Dahomey, the deified ancestors are arranged in families. Much like Dahomeans, Haitians believe that when ceremonies are held in honor of these ancestors, celebrants who experience spirit possession are the descendents of those ancestors. Spirit possessions, as well as the complex ritual details which accompany them, are also Dahomean in nature. For example, the drum rhythms and musical instruments used to incite possessions, the counterclockwise direction of the dances, the individual modes of behavior of persons possessed, are similar to those of Dahomey.

If so much of Dahomean religion has been retained in Vodou, it is because a large majority of slaves during the colonial period came from Dahomey. Although no one is sure exactly of how many Dahomeans came to Haiti, the fact is that their religious traditions dominated over those coming from other parts of Africa. Yet, as these slaves accustomed themselves to their new environment, the religious traditions which they brought with them came into contact with European Catholicism through the work of many French missionaries. These missionaries came to the colony with a sincere faith and pressing desire to reenact the miracle at Pentecost and to reach all men regardless of their race. For in Christ, there was to be no "Jew nor Greek, neither slave nor free." Filled with the unquestionable devotion to evangelize the world, the colonial church set in motion a number of events which were to leave their imprint on Haitian culture. Slaves were not allowed to hold religious meetings; the beating of the drums, and the religious and magical practices connected with Africa were made crimes. Offenders were tortured and often punished by death. The severity of such laws as the 1685 *Code Noir* (Cabon 1930, 33ff), which ordered all masters to have their slaves instructed and baptized "in the Catholic religion, apostolic and Roman" within eight days after their arrival, drove African rituals underground, and resulted in the nocturnal character of Vodou, a character which still persists today.

The hostility of the Roman clergy towards Vodou, coupled with the conviction that they had to respond to the Gospel call to evangelize the world, caused the slaves to overlay their African religious rites with a veneer of Roman Catholic rituals and symbols. In the contact between Africa and Europe which occurred on Haitian soil, and under the weight of the oppressive conditions of slavery, a process of acculturation began to take place. In responding to culture contact, the African slaves succeeded in achieving, contrary to the expectations of the European missionaries, what scholars have termed a syncretism, a merging of African patterns with the European religious traditions.

This religious syncretism developed into a system of correspondences between the largely Dahomean elements and Roman Catholicism of Europe. Such correspondences can be seen in several ways. First, the names of the saints were added to the already established African pantheons of deities. Such correspondences were not based upon the life of the saints, but upon certain symbolic accoutrements associated with them which corresponded to those found in the myths about the African deities. Thus, for example, the trickster deity Legba, who in Dahomean mythology is said to guard the crossroads of destiny and the entrances to the temples, becomes St. Peter because in Christian symbology St. Peter holds the key to the doors of heaven. Damballah, the Dahomean snake deity, becomes St. Patrick because of the Catholic legend about St. Patrick and the snakes in Ireland. Second, the correspondence between African practices and Catholicism can be seen in the hierarchical organization of the *hounfort*, where the behavior of the *prêt' savanne* (bushpriest) behavior corresponds to that of the Catholic priest in the Catholic ritual. The *prêt' savanne* is not the Vodou priest. His role is largely perfunctory in the *hounfort*. He brings to the Vodou rituals elements of the Catholic liturgy by baptizing persons, by blessing ritual paraphernalia, and by reciting prayers of the church whenever these are needed. The *prêt' savanne* is often an acolyte in the Catholic Church, one who has learned enough of the liturgy to become a symbol of the church in all Vodou meetings. In short, Vodouists see the *prêt' savanne* as a tangential figure, one who touches both the Roman Catholic Church and Vodou but is of central importance to neither. He is the symbol of the religious correspondence between the two religious systems.

Third, the system of correspondence between African and Catholic traditions can be seen in the use of Catholic objects such as the rosary, the chromolithographs of saints, and particularly the Christian cross. In Haiti, it is not at all unusual to find on the same *pé* (Vodou altar) Vodou ritual paraphernalia next to these Catholic objects. However, the juxtaposition of these objects should not lead anyone to think that Vodouists' use of these Catholic objects resembles those prescribed by the church. Just as the saints lose the personalities ascribed to them by Christian hagiology to acquire those of Dahomean deities, so too the symbolism of these objects takes on Dahomean characteristics. The Vodouists used them primarily as *gardes* (magical charms) to protect them against malevolent magic. Vodouists say that since these objects have been blessed by the Catholic priest, who is the direct representative of

God, the ruler of the universe and head of all the pantheons of *loas*, these objects contain a great deal more power than those Vodou objects initiated simply to the "service" of the *loas*. The most significant of these objects is the crucifix. Although crucifixes appear on every *pé*, their uses in the Vodou rituals have little to do with Christian iconology, but are interpreted in terms of Dahomean mythology. The cross is the most important symbol in Vodou. It represents the crossroads of human destiny as represented by Legba; it is the symbol of the avenue of communication between the profane world of men and the sacred world of the *loas*. Moreover, like the Dahomeans, Vodouists see in the cross the entire cosmos including its atmospheric, telluric, and subtelluric zones. It also represents the four cardinal points of the universe, points at which the *loas* are said to reside. In short, the cross is the central symbol in Vodou. It is the basis for Vodouists' *Weltanschauung*. They see it as the supporting framework around which the universe is constructed.

In view of the correspondence between the crucifix and the cross in Vodou the purpose of this paper is two-fold. It seeks by studying the use of the Christian cross in some of the Vodou ceremonies to understand the nature of the cosmos as Haitians see it. This understanding will permit some observations about Haitians' attitudes toward that cosmos. It will also show that, contrary to the opinions of many scholars who have written about Vodou, Haitians do not interpret the cross symbol in Christian terms, but, as with the saints of the church, in the light of African mythology, particularly that of Dahomey.

The Cross: Its Uses and Interpretations in Vodou

Like that of the Dahomeans (and most Africans), the supernatural world of Haitians is a rich, vast, and unexplored one. It is populated by more than one thousand *loas*, who are believed to reside variously in the silk-cotton tree on the rocky and steep footpaths on the mountain side, in the *reposois* (repositories)[1] in the courtyard of the *hounfort*, in the humblest thatched-roof hut, in the public fountains or the rivers where the people come to fetch water, or at the four cardinal points of the universe. By and large, Vodouists say that the permanent residence of the *loas* is in "Dahomey in Africa," which is an "island below the sea," more specifically in the mythological city *Ville-aux-Camps*. The events which occur at *Ville-aux-Camps* are no living person's business, though everyone has conjectures about them. According to the stories gathered during the course of this research, few living persons have ever been allowed to enter the city of the *loas*, those privileged few having been taken there "accidentally" by the *loas*.

Although few Vodouists have ever "visited" *Ville-aux-Camps*, most believe that it exists because they have established contact with it through the medium of the *houngan* in the *hounfort*. The *houngan* makes this contact at the outset of a Vodou ceremony by invoking Legba, the *loa* of the crossroads. In Haiti, as well as in Dahomey, the deities are said not to speak the same language as the human beings who serve them. Legba is the divine medium through whom men's requests and prayers can be channeled to the respective *loas*. Legba is the interlocutor, the interpreter, the principle of crossing and of communication with the divine world. Legba is the Hermes of the Vodou Olympus, the one who not only translates the supplications of men, but who is also the protector of the divine community. He is said to guard the doors to the gates of *Ville-aux-Camps*. During the Vodou ceremonies, he is invoked first because Vodouists say that he opens the *bayé* (gate) that separates the *loas* from their *serviteurs* (devotees). Legba is invoked in the rhythmic dances of the *yanvalou*[2] as follows:

> Atibo Legba, open the gates (to *Ville-aux-Camps*) for me
> Papa Legba, open the gates so that I might enter
> When I will go (to *Ville-aux-Camps*) I will salute the *loas*
> Vodun Legba, open the gates for me,
> When I will go in, I will thank the *loas*.

As this song indicates, the invocation of Legba is an important part of the Vodou ceremony. As the *Société* (devotees) dances and sings it, the *houngan* solemnly traces Legba's *vèvè*,[3] an act which is said to open the gates to *Ville-aux-Camps* and thus to permit the passage of the *serviteurs*' supplications to the divine world. The focal image of this *vévé* is Legba's symbol, the cross. In its form, this cross resembles the Christian cross but, in its significance, it is entirely African.

The symbol of the cross is central to Vodou's ceremonies. Indeed, Vodouists revere it wherever they encounter it: not only when it is traced on the *pérystile's*[4] floor, but wherever they encounter it in their daily lives. For example, in what might be called in this paper a rite of territorial passage, Vodouists who pass a sacred edifice (a Catholic church, a school, or a hospital) usually cross themselves reverently. To them, the Christian cross which dominates the building symbolizes Legba, the medium through which contact can be established with the world of the *loas*.

Because Vodouists usually appear to revere the Catholic cross and because the cross image has such a symbolic significance in Vodou meetings, many Haitians as well as foreign scholars who have written about Vodou have argued that Vodou crosses are borrowings from Roman Catholicism. While it is true that the Catholic crosses in Haiti may well have been invested with Vodou meaning, the cross symbol in Vodou does not originate in Catholicism but in African mythology. In Dahomey, for example, the universe is conceived as a sphere transected by two mutually perpendicular and intersecting planes, which, perceived in a cross-section of the sphere, represent the arms of a cross (Mercier 1954, 219–22, 224–25). These crossed planes are conceived to provide the framework and supporting axes of the sphere. In the myths, this cross resulted from the movement of Mawu (the Dahomean Godhead) to the four cardinal points of the universe while she fashioned it. The Dahomean creation myth compares the universe to two halves of a calabash whose edges match perfectly. The plane along which the two halves are joined together constitutes the plane of the horizon. Dahomean mythology also conceives of two different crosses in relation to the plane of the horizon: first, it conceives of the four cardinal points of an earth which stretches flat along the plane of the horizon; second, it conceives of a vertical plane intersecting the mid-point of the horizontal plane, thus forming a cross with the horizontal plane. Each end of that plane penetrates the walls of the calabash. The endpoints of the cross on the horizontal plane which penetrates the calabash represent the four cardinal points of the universe. Similarly, the end-points of the cross which stretches flat on the surface of the earth mark the four cardinal points of the earth: west, east, north, and south.

Dahomeans also believe that the calabash which constitutes the cosmic sphere is surrounded by a large calabash welded in the same manner as the small one. Beyond the larger calabash is the abode of the deities. The calabashes are separated by a large body of rain water. Since the small calabash is mobile, it is said that the "little calabash floats in the larger one" (Mercier 1954, 220). The small calabash contains the visible world with its elements: earth, water, fire, and air. In both Dahomean and Haitian mythologies, the earth is conceived to be a flat disk surrounded by and floating on water. This is why, Dahomeans and Haitians say, one finds water when one digs deeply into the earth.

According to Dahomean mythology, when Mawu Lisa arranged the universe from pre-existing materials, she traveled throughout the universe and stopped at the "four quarters of space," which correspond to the four cardinal points of the earth. At each halt, she gathered materials with which she formed each half of the calabash, solidifying the dirt from which the earth is composed and determining the location of the waters. The path of Mawu's primordial movements formed a cross whose lines extend from west to east and from north to south. Dahomeans say that when Mawu created men, she distributed them at the four cardinal points of the universe. At each point, she spent one day. On the first day, *ajaxi*, she fashioned men in the west; on the second day, *miõxi*, she created men in the east; on the third day, *odokiwi*, she made men in the north; and the fourth day, *zobódo*, she placed them in the south. Dahomeans believe that each act of creation at the four cardinal points of the universe resulted in the formation of four countries: *Aja* on *ajaxi, oyo* on *miõxi, Ke* on *odokiwi* and *Hū* on *zobódo*. Dahomeans believe that in the south, *Hū* is Dahomey because it was created last, and it is said to be "called to a noble destiny" (Mercier 1954, 224–25).

The cross represents the path not only of Mawu's travels but of those of other minor deities who were responsible for completing the act of creation. According to the Dahomean story, Gu, the giver of civilization and art, retraced Mawu's steps, moving from west to east, and from north to south. A similar pattern occurred when the different sibs were founded.

The Africanness of this symbol is made clear by the fact that the cross is significant not only in Dahomean worldview but in other parts of Africa as well. Germaine Dieterlen (1950) points out that the figure of the cross is an important symbol in the Bambara worldview (Dieterlen 1930, 2–33). Among the Bambara the world is believed to have been created by Faro. Unlike Mawu, Faro is not the arranger of the world but its creator; he fashions it not out of pre-existing materials, but out of a void, or *glã* (Dieterlen 1950, 2–3, footnote 5). The Bambara's concept of the universe is founded upon two basic cosmological principles: the eternal vibration of matter and the movement of the universe as a whole. According to the myth of creation, the original germ of the world came out of *glã*, or emptiness, enveloped by a sheath. In the beginning, *glã* "spoke loudly" and the sound of its voice caused the germ to vibrate and to move in the form of a cross. Out of this vibration the germinating principles within *glã* produced its twin. From this primordial pair of "energy principles" flowed the entire universe. According to the myth, deities, human beings, animals, and matter emerged out of *glã* and reached the uttermost confines of the universe. Faro, the Godhead, later arranged them at the four corners of the universe. According to Dieterlen, the Bambara based their worldview on two active principles: the phenomenon of twin-ness (the principle of efficient cause in matter and biological reproduction in the deities and men) deriving out of a primordial energy-force, and the sign of the cross (connecting the four cardinal points of the universe), which establishes cosmic order in the universe (cf. Griaule and Dieterlen 1954, 84).

The cross is thus the metaphysical axis around which the universe is constructed. On the one hand, it holds the entire universe in equilibrium. On the other, it provides the path for the infinite extension of the universe by the continual progression of matter. Dieterlen adds that among the Bambara the symbol of the cross appears on most of the sacred ritual objects and on many of the cooking utensils. It is also drawn on the ground at religious ceremonies and serves to frame a game of a quasi-ritualistic nature called *sumangolo. Sumangolo*, observes Dieterlen, is a game to which "the Bambaras of both sexes abandon themselves over" to contribute to what the Bambara call "la marche du monde" (the way of the world) (Dieterlen 1950, 31–33).

An examination of the cross symbol in Vodou will make it clear that both its ritual function and its theological significance derive from African mythology and not from Roman Catholicism. Like the Dahomeans, Vodouists see in it a cosmographic image which symbolizes their *Weltanschauung*. It is used in religious ceremonies which establish contact between the world of men and that of the *loas*. The cross represents both the fact of communication between the two worlds and the nature of the difference between their modes of reality. In the first of these symbolisms, the horizontal line of the cross represents the world of the living, and the vertical line the medium of communication with the abyss. In the second symbolism, the vertical line represents the other world itself, its verticality indicating that the reality there is the inverse of the reality in the world of the living. This latter symbolism becomes clear when one realizes that in the Vodou worldview the metaphysical world of the *loas* is not a vague and mystical one but a cosmic mirror which inverts the images of the world of the living in reflecting them. The *loas* are often referred to as mirrored images reflecting the deportment and the personalities of men. For example, when the *houngan* addresses or calls upon them, he does so in terms which describe them as reflectors of man: Loco-mirroi, Agassou-Dos-mirroi.[5] Moreover, during Vodou ceremonies the mirroring of the image of the world of the living is also symbolized by a number of ritual observances. When the *houngan* greets his *laplace* or assistant, the two face each other holding each other's hands and perform a number of turns—first clockwise, then counterclockwise—to represent the sights of the profane world of men as reflected in the cosmic mirror of the *loas*. A mother who comes to present the new-born child to the *loas* holds him on her left arm. As already noted, the ritual dance in the *pérystile* also revolves in a counterclockwise motion around the *poteau mitan* (center pole).

Not only is the physical detail of the world inverted in the cosmic mirror of the sacred world, but so too is the order of progression of time. The *houngan* who is about to begin a ceremony often enters the *pérystile* by emerging out of the door of the *bagui* (sacred chamber of the temple) in a backward motion (Deren 1970, 35).

This reverse movement symbolizes the retrogression through time to the first man, to the unmanifested primordial man, and finally to *Bon Dieu* himself, the Godhead who created the universe.

The opening of the Vodou ceremony represents the sudden halt of profane time. The entire ceremony in which a large number of *loas* "mount"[6] their *serviteurs* (devotees) is the archetypal reconstruction of the cosmos *ab origine*, and as the deities appear in succession, the *serviteurs* participate in the exploits of the creation of the world executed by *Bon Dieu* who fashioned the universe as a reflection of his own world. At the start of a Vodou ceremony, cosmogony is reactualized; the *serviteur's* possession symbolizes the recreation of the Vodou pantheons; the world of the living is refashioned, and the individual, as well as his community, is born anew. The repetition of the archetype represents the temporary suspension of profane time.

The principle of inversion and retrogression is fundamental to the Vodouist's mythology as well as to his rituals. Hence, in Vodou, the relationship between the cosmic mirror and the mundane reality which it reflects takes the cosmographic form of the cross. In the cross, Vodouists see not only the earth's surface as comprehended by the four cardinal points of the universe, but also the intersection of the two worlds, the world of men as symbolized by the horizontal line, and *Ville-aux-Camps* as represented by the vertical. The foot of this vertical line plunges into the waters of the abyss to the cosmic mirror where the *loas* reside. Here, on the island below the sea is Dahomey (or *Ville-aux-Camps*), the legendary place of racial origin. The point of contact at which, in Vodou ceremonies, the two lines intersect is the pivotal "zero-point" in the crossing of two worlds. It is a point of contact at which profane existence, including time, stops and sacred beings from the island below the sea invade the *pérystile*.

The cross is therefore the most important ritualistic symbol in Vodou. It appears wherever communication or traffic between the two worlds occurs. The following examples will indicate some idea of its pervasiveness in Haiti.

Perhaps the most glaring example is to be seen in the structure of the *pérystile*. In Vodou, the *pérystile* is a microcosmic representation of the universe. The four poles sustaining the structure symbolize the four cardinal points of the universe, covered by an overarching roof which represents the cosmic vault above the earth. Like the horizontal line of the cross, the floor of the *pérystile* symbolizes the world of man while the vertical *poteau mitan*[7] represents the *axis mundi*, the avenue of communication between the profane and the sacred world. Although the vertical reach of the *poteau mitan* appears to be limited by the *pérystile's* structure, mythologically its foot is conceived to plunge into the cosmic mirror. The point at which the *poteau mitan* enters the floor of the *pérystile* symbolizes the zero-point. In the physical structure of the *pérystile*, the point of crossing is usually marked by a round platform built around the *poteau mitan*. During ceremonies, the *poteau mitan* becomes charged with the power of the *loas*. Hence, before tracing the *vèvès* the *houngan* touches the pole, a ritual act which empowers him to summon the *loas* into the *pérystile*. Thereafter, like the *poteau mitan*, the *houngan's* body becomes in itself the source of power, a repetition of the microcosmic symbol, a moving embodier of the vertical axis around which the universe revolves.

The cross structure of the *pérystile* repeats itself in the geometric symbols, or *vèvès*, which are central to the rituals. The drawing of these *vèvès* reflects the concept of inverted symmetry and is characterized by a series of motions involving doing-and-undoing.[8] As already noted, the *vèvès* are drawn on the floor of the *pérystile* by the *houngan* at the beginning of every ceremony in order to invoke the *loas* to manifest themselves in the bodies of their *serviteurs* during the ceremony. Since each *loa* has his *vèvè*, the area where it is traced becomes consecrated to him. In drawing the *vèvès*, the *houngan* traces the cross first. Holding some flour between the thumb and the forefinger of his right hand, he meticulously draws each line, applying the method of doing-and-undoing. First, he sifts the flour, moving his hand from left to right and then from right to left, from top to bottom and then from bottom to top. The drawing of the cross completed, the *houngan* draws the various symbols which represent the personality ascribed to the *loa*. Following the same pattern of doing-and-undoing, each stroke of the hand, each circle drawn upon the cross or around it is indicative of the forward and reverse, clockwise and counterclockwise motions of the world and its reflection.

The complex representation of the deity's personality appears to hang on the two intersecting lines of the cross. Once finished, the *vèvè* appears to the worshipers as both static and dynamic. As principal posts of a building support their trestle, the lines of the cross provide a structure on which these images hang in equilibrium. But also, like a revolving door, these geometric images seem to spin (in both clockwise and counterclockwise motions) as if striving to throw off the images to the outer limits of the earth's surface. Both of these semblances are appropriate for, although the lines actually lie flat on the floor of the *pérystle*, the Vodouists see the horizontal line of the *vèvès* as representing the secular world, while the vertical line reaches the cosmic realm of the *loas*, and plunges into the waters of the abyss.[9]

Sometimes *vèvès* are elaborate and complicated symbols, the white lines of which appear as fragile laces. This sense of fragility is heightened when one considers that the slightest breeze can blow away the light flour which composes its lines (Deren 1970, 204). Yet, these fragile symbols communicate profound mythological significance. For through the vertical line of the cross, the *loas* are summoned to visit their *serviteurs* in the *pérystile*. The point at which the two lines intersect indicates the central location at which human dialogues no longer follow the horizontal path of mundane communication, but plunge sharply toward the world of the *loas*, burying with them the words of the supplicative songs which accompany the tracing of the *vèvès* and are forced downward. As if amplified in weight a thousand times, each grain of flour which falls to the ground raps the sacred portals of the *loas* to signal them to possess their *serviteurs*.

Possession indicates to the *société* that, in spite of the remoteness of his permanent home, the *loa* is also one who comforts his *serviteurs* in the anxieties and the defeats of their lives and that, at the root of the cosmos, goodness always endures. Even with their poverty, their hunger, their failures, peasants know that they are rewarded by the cosmic power of the loas, who are capable of bestowing on their *serviteurs* the hope of a bright future. The intimate relationship that a possessed *serviteur* establishes with a *loa* through the meditation of the *vèvès* accords him many benefits. The *loa* who mounts a *serviteur* soothes his particular fears and helps him in his personal losses.

The presence of the *loa* in the body of a *serviteur* is a nonmaterial achievement which symbolizes the unquestionable devotion of the *loas* to intervene in the profane activities of men. During the period of possession, the *serviteur*, like the *houngan* before him, embodies the cross symbol of the *vèvès*, the zero-point of contact between the sacred and the profane world. He becomes a medium whose feet are planted in the sacred mirror and whose body is the vertical line whereby the revitalizing forces of the universe flow to the *société*.

When a *loa* invades the body of a *serviteur*, he greets the *société* with a series of movements which reflect the same pattern of doing and undoing seen in the drawing of the *vèvès*. The possessed *serviteur* greets the *société* with a double handshake, with the right hand first and then the left. The crossing of the hands symbolizes the crossing of two worlds—the right representing the horizontal line of the cross, the world of men, and the left representing the vertical line of the deities. Because the left hand of a possessed *serviteur* represents the intervention of *Ville-aux-Camps* in the world of the living, Vodouists believe that it is charged with the psychic power of the *loas*. When the left hand of a possessed *serviteur* grips the left hand of another, the power of the *loa* which mounts the possessed *serviteur* is transmitted into the body of the other to possess him. While the possessed holds another with his left hand the power of the *loa* spins the other in a series of counterclockwise movements.

Summary

In examining the use of the cross in Vodou, we can isolate two main ideas: First, as among the Bambara, the tracing of the *vèvès* at Vodou ceremonies can be interpreted as the symbolic recreation of the universe. In this sense, by the intricate tracery of the *vèvès*, and the intervention of the loas, Vodouists not only relive the cosmological archetype in their retrogression through time, but by being possessed, they also relive the recreation of the world. "Creation" for Vodouists is not a static event, but a dynamic process which, as for the Bambara, is an ever-recurring "marche du monde." When Vodouists observe the universe, they see both man and the substances which surround him as part of nature. In the context of this dynamic "marche du monde," Haitians also see two different categories of change in its operation: first, the degenerative changes in persons and in substances which eventually decay and die, and second, certain patterns of recurrence—the cycle of generations, the cycle of seasons, the recurrent movement of the astral bodies. These are the pervasive and persistent principles upon which the universe operates. Vodouists see these principles as immortal and, therefore, as belonging to an order of existence superior to matter. It is these principles that they identify as the *loas*. Hence, to Vodouists, the cross is not only that medium of communication between man and the *loas*, but it is the symbol through which man can relate to the cosmological principles which the *loas* personalize.

Second, the *Weltanschauung* of Vodouists corresponds to that of the Dahomeans. The universe envisioned by Vodouists can be described in two ways. On the one hand, in a macrocosmic dimension, Vodouists conceive of space in the universe as limited by four cardinal points. The points are connected by lines which take the form of a cross, a metaphysical axis which supports the entire structure of the universe. On the other hand, in a microcosmic dimension, the *pérystile* is the symbol of the universe. The space within it is also limited by the four poles which support the structure. The metaphysical line which joins these four poles crosses the floor of the *pérystile* to form a horizontal cross, which is limited by the area of the *pérystile*. Moreover, the vertical *poteau mitan* intersects the floor of the *pérystile* and mythologically supports the entire structure of the universe.

As the cosmic reflection of the world of man, *Ville-aux-Camps* is conceived by Vodouists to be structured much like the *pérystile*, for the *poteau mitan* which traverses the *pérystile* also traverses the sacred world of the *loas*. As already noted, during the Vodou ceremonies, the *vèvès* and the bodies of possessed individuals become a further microcosmic reduction of the *poteau mitan*. Hence, the cross symbol in Vodou cannot be seen as representing a symbolic relationship between Roman Catholicism and Vodou but indicates the retention of African religious elements on Haitian soil. Indeed, for Vodouists, the Christian cross which dominates a Catholic building indicates the zero-point at which the two worlds bisect one another. The steeple of the church, for example, is analogous to the *poteau mitan*. Consequently, when Vodouists pass in front of churches, they first cross themselves at mid-point to symbolize their territorial passage through sacred ground. The paths trod by their feet represent the horizontal plane, the profane world of men. Such territorial passage must be completed without intermittent rest in order to permit both lines of the cross to intersect one another. According to Vodouists, it is therefore a bad omen to stand in front of churches to talk because *Bon Dieu* (the Godhead) may be angered and may inflict illness and even death on them.

While it may be true that the African symbol of the cross among the slaves could have disappeared (as did many aspects of African religion) if it had not been reinforced by the parallel Catholic symbol during the colonial period, still in response to the early missionaries' catechizing efforts, the slaves learned to interpret the Catholic cross in terms of their own African religious traditions.

Notes

1. Rectangular area in the courtyard of the hounfort where the *loas* are said to reside momentarily.
2. The yanvalou is a ritual dance form dedicated to the snake *loa* Damballah.
3. Cabbala-like geometric designs representing the function of the *loas/lwa.*
4. Section of the hounfort in which ceremonies are held. Each *loa* has a separate design. They are traced at each ceremony usually before possession occurs. The origin of these designs has not been clearly established although some students (Maximilien, Louis, Le Vaudou haitien [Port-au-Prince: Imprimerie de l'Etat, 1954?]; Maya Deren, *Divine Horsemen: The Voodoo Gods in Haiti* [New York: Delta, 1970]) of Vodun have conjectured about their pre-Columbian Indian origin.
5. Literally "Agassou in the mirror's surface, Loco in the mirror."
6. In Vodou, the possessed person is compared to a horse. Vodounists say that he is ridden by a *loa.*
7. The poteau mitan is a pole at the center of the pérystile. It serves as the central channel of communication between the sacred world of the *loas* and ancestors and the profane world

of the living. During a ceremony, participants dance in a counterclockwise motion around the poteau mitan.

8. As described below, doing-and-undoing refers to the manner in which the person drawing the vèvès moves his hands from left to right, from right to left, from top to bottom, and from bottom to top. Every movement is followed by its inverse.
9. Although the houngan traces the *vèvès* on the floor of the *pérystile*, mythologically it is not horizontal but vertical.

References

Cabon, Père Adolphe. 1930. *Notes sur l'histoire religieuse d'Haiti, de la révolution au concordat 1789–1860*. Port-au-Prince: Imprimerie Compiègne.

Charlevois, Xavier de. 1730–1731. *Histoire l'Ile espagnole ou S. Domingue*. 2 vols. Paris: Hippolyte Louis Guérin.

Deren, Maya. 1970. *Divine horsemen: The Voodoo gods in Haiti*. New York: Delta.

Desmangles, Leslie G. 1975. God in Haitian Vodun: A case in cultural symbiosis. PhD dissertation, Philadelphia, Temple University.

Dieterlen, Germaine. 1950. *Essai sur la religion Bambara*. Paris: Presses universitaires de France.

Griaule, M. and G. Dieterlen. 1954. The dogon. In *African Worlds: Studies in the Cosmological Ideas and Social Values of African Peoples*, ed. Daryll Forde, 83–110. London: Oxford University Press.

Herskovits, Melville. 1971. *Life in a Haitian valley*. New York: Doubleday.

Jahn, Jaheinz. 1961. *Muntu: An outline of the new african culture*. New York: Grove.

Maximilien, Louis. 1954? *Le Vaudou haitien*. Port-au-Prince: Imprimerie de l'Etat.

Mercier, P. 1954. The Fon of Dahomey. In *African Worlds*, ed. Daryll Forde, 210–234. London: Oxford.

Price-Mars, Jean. 1928. *Ainsi parla l'oncle*. Port-au-Prince: Imprimerie Compiègne.

Saint-Méry, Moreau. 1797. *Description topographique, civile et historique de la Partie française de St. Domingue*. 3 vols. Philadelphia (no publisher given).

Chapter 4

The Song of Freedom: Vodou, Conscientization, and Popular Culture in Haiti

Gerdès Fleurant

The themes of justice, liberation, peace, progress, and unity dominate the music of the popular culture today in Haiti and its Diaspora, now renamed the Tenth Department. While the people of Haiti have been in the forefront of the world resistance movement against colonialism and political oppression for the last two centuries, and Haitians are known throughout the world as a fighting people (Charlie Mingus, the great African-American jazz musician recorded in 1957 a composition titled "Haitian Fight Song"), it is only in the past ten years that its popular music has explicitly articulated the themes that in the 1970s were referred to as "Kilti libète" or Freedom Culture. This movement that is not limited to music, for one finds its counterparts in the dance, drama, literature, and the plastic arts, results in part from a long accumulation and internalization of resistance patterns, and the world progressive movement that came of age in the wake of the 1960s social consciousness revolution spearheaded by the Civil Rights Movement in the United States. This chapter, an exploration into the Haitian song of freedom from its roots in Bwa Kayiman to Boukman Eksperyans and beyond, will focus on the role of music in African and African-derived cultures, with specific emphasis on the Vodun. It will further highlight the centrality of Haiti, its Vodun and music, in the contemporary world movement of oppressed people's cultural affirmation. Finally, the dialectical relation between the various elements of the world movement of Conscientization (Freire 1972; Nkrumah 1970; Fanon 1959; John XXIII, 1958–1963; and U.S. Civil Rights) will be elucidated to explain the role of the arts, music in particular, in the future democratization of our planet's socioeconomic/political structures, a movement I call "Humanocentrism."

Bwa Kayiman and the Haitian Revolution

The European encounter with the New World through the adventures of Christopher Columbus and the search for new routes for commerce resulted in the enslavement of millions of souls from Africa. Slavery, a practice that supported the world economy for more than three centuries (1492–1888), reached its peak and at the same time met with its demise at the hands of the African rebels, on the island of Saint Domingue, the former nation of Haiti.

Saint Domingue, France's prized colony in the Americas, flourished so well that by 1791, the date of the Haitian Revolution that began with the politico-religious congress of Bwa Kayiman, it counted a population of 700,000 Africans, about 40,000 French persons, and an equal number of mulattos or freed persons, from the union of the white men and Black women. According to James G. Leyburn, in 1789, two-thirds of France's commercial interests were centered in Saint Domingue, and the territory's combined export-import business was valued at more than $140,000,000. The economic importance of the island is further reflected in the fact that "its sugar, coffee, indigo and cotton supplied the home market and employed, in prosperous years, more than 700 ocean-going vessels, with as many as 80,000 seamen" (1966:15). The conflicts that resulted from such an accumulation of wealth produced through forced labor accounted for the events, known in historical annals as the Haitian Revolution, that shook the colony between 1791 and 1803. Indeed, the enslaved Africans who were brought to the colony as early as 1512 to work in the agricultural plantations, after a period of marronnage (guerilla warfare in the mountains) and sustained organization, succeeded in overthrowing the hated system of human slavery.

Central to the success of the Haitian Revolution was the faith of the captives who revolted in their culture and way of life, the Vodun, a comprehensive socioeconomic system of universal knowledge. Issued from Africa, a continent whose culture and philosophical vision do riot in order to distinguish between the sacred and the secular, the organizers of the revolt gathered in the area known as Bwa Kayiman during the night of August 14, 1791, to consult the spirits of their ancestors or the *lwa* of Vodun. It was there that they intoned their first revolutionary hymns and danced their first liberative spiritual entities: Legba, Danabla, Ezili, Ogoun, Simbi. It was there that they formulated a comprehensive approach to the arts as a liberating factor in human affairs. It was there, at the Bwa Kayiman religious congress, that they consecrated the role of culture as an indispensable component of political action. Thus the roots of the song of freedom lie deep in the memory of Bwa Kayiman and beyond, in Africa, the spiritual home of the ancestors. Today's affirmation songs of Boukman Eksperyans, Boukan Ginen, RAM (Roots Music Group), Foulah, and many other Haitian musical groups that promote the traditional culture under the name of "mizik rasin" logically embraced the spiritual experience of their forebears at Bwa Kayiman, and passed through the cultural movements of rehabilitation such as Indigenism and Negritude.

Indigenism and Negritude

The roots of the Haitian song of freedom also go back to the Indigenist movement of the 1920s and Negritude of the 1930s. Indigenism emerged as a reaction to the first American Occupation of Haiti (1915–1934), which crushed the Haitian elite's aspirations to Francophile status when American troops, mostly from the South of the United States, discovered their aspirations; the elite was, in the words of William Jennings Bryan, "niggers speaking French." After the defeat in 1919 of Charlemagne Péralte and the Cacos, the group of irregulars considered in Haitian history as nationalist freedom fighters that he led, the resistance shifted to the realm of the pen. Haitian authors began to search for their African heritage through the study of folklore and peasant life.

Indigenism was a pivotal movement in the history of Haiti that, from the colonial period to independence in 1804 to the first American Occupation in 1915–1934, consisted in a succession of fratricidal strives among different social groupings: Blacks against mulattos, rich against poor, and Catholics or Protestants against Vodunists, capped by a rejection of their African roots in favor of European affiliations. Indigenism was pivotal not only in raising the consciousness of the Haitian elite as to their African cultural heritage, but the movement also inspired the birth of Negritude, its better known counterpart in the annals of Pan-Africanism, the umbrella movement grounded in the concept of unity of aspirations of all peoples of African descent. Aimé Césaire, the father of Negritude, attested to this when he commented that Negritude was born in Haiti.

Indigenism influenced the Harlem Renaissance in the United States as well. It may be appropriate to mention here the fact that Haiti, particularly the example of the Haitian Revolution, was the single most important event in the history of the Black World during all of the nineteenth century and part of the twentieth century. Haiti inspired the three major slave revolts in the South of the United States led by Gabriel Prosser in 1800, Denmark Vesey in 1822, and Nat Turner in 1831 (Hunt 1988:107–146). The cultural and literary movements, such as Indigenism, the Harlem Renaissance, and Negritude can be better understood when placed in the context of the larger struggle of people of African descent for both physical and mental liberation. It is instructive to note that W. E. B. DuBois, one of the foremost intellectuals and forefathers of Pan-Africanism, understood well the issues when, in his book *Souls of Black Folk* (1969–1903), he cast the struggle in the following terms: "[T]he problem of the twentieth century is the problem of the color line—the relation of the darker to the lighter races of men in Asia, Africa, in America and the Islands of the sea" (54). DuBois's incisive comment has already anticipated a central component of the present day North/South debate, which calls attention not only to the economic disparity between rich and poor nations, but also the fact that the nations of the southern hemisphere are inhabited by darker hued people whose histories are grounded in colonialism. To DuBois, the unity of the struggle suggested a conceptual framework that englobed the darker races of Africa, Asia, and the islands of the Caribbean. And at the center of this "prise de conscience" were the people of

Haiti and their determination to liberate themselves from the shackles of colonialism. Thus, it is no accident that one of the key works of Indigenism, Jacques Roumain's *Gouverneurs de la Rosée* (1944) was translated into English, under the title *Masters of the Dew* (1978), by two celebrated poets and writers of the Harlem Renaissance, Langston Hughes and Mercer Cook. Hughes and Cook were attracted to the liberation and self-reliance theme that runs through the novel in which water as a symbol of life figures prominently. *Masters of the Dew* consecrated the triumph of science grounded in cultural practices over ignorance, greed, and hatred. Manuel, the hero of the story, although a nonbeliever in Vodun, found himself dancing with all of his soul at the ritual offered in honor of the *lwa* at his parents' compound to celebrate his return from 15 years of absence. His ability to find the water in a drought-stricken village and to search for a way to get it to the people or, better, to teach the people to get it for themselves (for he advised a *konbit* or cooperative work) put him in direct opposition with the oppressive regional political establishment forces; so in the end he was assassinated, in fact sacrificed one could say, but not before he laid the seed of life in his wife and companion Annaise, a most tangible symbol of the continuity of the race. *Masters of the Dew* symbolizes the determination of the common people to fight together for their total liberation, and the book stands as proof that humanity's ontological vocation, in the words of Paulo Freire, is to be free. The literary aspirations of the elite found its most eloquent expression in the 1946 bourgeois social revolution, a culmination of the grievances of the past century where the peasant, the urban proletariat, and the vodunist have remained unheeded. If Indigenism and Negritude spawned a literary movement, "1946," as it is known in Haitian history gave birth to a cultural renaissance that found its way in music and dance, all grounded in the traditional religion of the people, the Vodun.

The 1946 Bourgeois Social Revolution

The 1946 bourgeois social revolution was a logical continuation of the previous movements that revindicated a better distribution of the nation's resources among its citizens. The first American Occupation (1915–1934), which occurred at a time when the northern power practiced racial discrimination, left Haitian society with a structure that favored the minority (mulatto) over the majority (darker skin people). Thus the 1946 social movement, whose aim was the reversal of that state of affair, championed to some extent the cause of the peasantry and the urban proletariat and promoted the candidacies of President Dumarsais Estime and the populist Deputy Daniel Fignolé, whose supporters were known as "rouleau compresseur" or bulldozer. Intellectual Black leaders such as Lonmer-Denis, François Duvalier, Mesmin Gabriel, René Piquion, and Jean-Baptiste Romain rose up to promote the ideologies of the previous generation, also known as the "Griot School," which they convened into their own brand of Indigenism, better explained under the term of "Noirisme." Noirisme crudely pictured the Haitian social situation in terms of antagonism between Blacks and mulattos without taking into account the dynamism of class, a tendency that aimed to absolve the brutal oppressions of dictators like Paul E. Magloire

(1950–1956) and the infamous Duvaliers (1957–1986). Although the 1946 social movement was fraught with internal contradictions, it nonetheless managed to spawn a cultural renaissance in the arts, folklore, and music, a point worthy of our attention.

The seeds of the cultural movement that flourished after the 1946 social revolution were sown as early as the 1930s and 1940s when musical groups promoted by families led by patriarchs such as Felix Guignard, and Lincee and Fabre Duroseau, emerged to play the Haitian meringue, based on the kongo rhythm whose roots are lodged in the ritual setting. In reaction to the anti-superstitious campaign of the Elie Lescot government that, in 1942, banned Vodun ceremonies, Mardi Gras and Rara festivals, under the pretext that they were relics of savagery inherited from Africa, the Jazz des Jeunes was founded in 1943 to promote the music and folklore of Haiti through the use of its traditional rhythms like the *yanvalou*, *ibo*, and *petro*. The Jazz des Jeunes, now an institution in Haitian popular music and culture, celebrated its fiftieth anniversary in 1993. For a long time, they were one of the few groups that played the traditional rhythms of Haiti, a distinction that won them the scorn of the Fundamentalist Protestant missions.

Lina Mathon Blanchet, a classically trained pianist, played a significant role, although inadvertently, in the promotion of Haitian folklore. According to African-American dancer/choreographer Lavinia Williams Yarborough, it was a group of German artists who suggested to Blanchet that the chorus she directed sing some folkloric songs. She refused, hesitated, and then accepted and went on to produce some "contes Haïtiens" with songs to the delight of the listeners. In 1941, Blanchet brought her Legba Singers to Washington, DC, for the Pan-American Conference, and later in 1951, she produced "Haïti Chante" again in Washington, DC (Williams Yarborough 1964).

But two events, the end of World War II and the opening of the 1949 Bi-Centennial Exhibition in Port-au-Prince, contributed to spark a tourist explosion, so to speak, which in turn led to a renaissance in the arts and folklore of Haiti; in 1949, a brand new city was built on the seashore of the capital to celebrate the two hundredth anniversary of the founding of Port-au-Prince. It was in this brand new city that an open air theater, the "Theatre de Verdure" was located. The theater became the center of the movement for the promotion and production of dance, folklore, and traditional music of Haiti. Besides the Jazz des Jeunes, mentioned earlier, a series of dance and musical troupes, with names like "Dambalah Dance Troupe," "Aida Troupe," "Erzuli Dancers," "Dantor Dancers," and "Macaya Folk Group," emerged on the scene to perform the rhythms of the Vodun rites of Haiti (Yarborough 1964). To these, we must add the famous "Choeur Simidor" directed by the late musician Ferrère Laguerre, and the well-known "Choeur Michel Déjean." In addition to the elegant kongo, itself the basis for most Haitian popular music and dance, those artistic groups experimented with considerable success with arrangements of rhythms like *yanvalou*, *ibo*, *petro*, and *nago*, all of them belonging to the sacred repertoire of the people's traditional religion. This period saw the rise of the incomparable singer Lumane Casimir whose voice is yet to be equaled in the annals of Haitian musical arts. Her renditions of songs like "Papa Gege Bel Gason," "Isit an Aym," and many others, were the delight of tourists, politicians, and the general public alike. Yet, like Billy Holliday and Mozart, Casimir died prematurely in abject poverty and obscurity. Her memory is yet to be rehabilitated by a Haitian government, and we are grateful to poet Jean-Claude Maroneau and singer Carole "Maroule" Demesmin for their tribute to this seminal artist (Carole Maroulé, Fleetwood CDM 1277 ca. 1983).

The 1950s and early 1960s brought about the use of folklore at official events such as the famous May 18 school children parade, held at the Port-au-Prince common, known as the "Champs de Mars," in commemoration of the creation of Haiti's flag on that date in 1803. Thousands of high school children danced along with their monitors to the music of the late Antalcidas Murat, titled "Sur les Rives de l'Arubonite," composed in *yanvalou* and kongo franc rhythms. Murat was the arranger and the musical genius behind the Jazz des Jeunes, and this writer's first music teacher, as well as that of a whole generation of Haitian musicians. Other musical geniuses worthy of mention are guitarist Frantz Casséus, and singer/scholar Emerante de Pradine who extensively use Vodun material as the basis for their musical selections. De Pradine, a member of the elite, shocked her entourage when she openly embraced the Vodun in her singing. She went on, however, to record some excellent songs with the Choeur Michel Déjean (1953) and, as important, to give birth to Richard A. Morse, one of the founders and promoters of the 1990s Haitian Root Music (to be discussed later).

Although the post-1946 artistic renaissance followed as a direct result of the Indigenist movement, the emphasis was still folkloric, literary, and elitist, judging by the names of many of its protagonists who were poets and writers. It is important, however, to mention some of the most prominent, who, in fact, were the springboard for the more progressive movements of later decades: Lorimer Denis, Emmanuel C. Paul, Lamartinière Honorat, Jean-Léon Destiné, Katherine Dunham, Lma Mathon Blanchet, Robert Baussan, Dumarsais Estimé (president of Haiti, 1946–1950), Félix Morisseau-Leroy (the foremost advocate of Creole culture, which he conceptualizes as Creolophony, in opposition to Francophony), Jean Brierre, Théodore Beaubrun (Haiti's foremost comedian and the father of Arnold, Daniel, and Theodore Beaubrun of Boukman Eksperyans), Pierre Blain, Lavinia Williams Yarborough, Emérante de Pradine, Alphonse Cimber (one of the leading master drummers of Haitian ritual), Antoine Salgado, Roger E. Savain, Marcel Sylvain, Renee St. Aude (one of the founders and powers behind the Jazz des Jeune), Louinès Louisnis, Joe Archer, André Narcisse, and André Germain. Moreover, this list is far from being complete, but it gives us an idea of the epoch and the orientation of the men and women who preceded the Rasin or Root Music Movement of today.

C. L. R. James's (author of *The Black Jacobins*) emphasis on a cumulative approach to social revolution should clarify for us the importance and role of the intellectuals in the evolution of the Haitian masses in their march to conscientization. The work of the Indigenists, like Jean Price-Mars and Jacques Roumain, the founder of the Bureau of Ethnology (1941), whose writings began to challenge some of the elite's preconceived notions about Haitian culture and Vodun, made it possible for 1946 to occur and the generation of the 1970s to emerge with the concept of "Kilti libète" (Freedom Culture).

The Freedom Culture of the 1970s

While the 1946 bourgeois social revolution occurred as a fulfillment of the ideologies of the previous generation, in particular, Indigenism, the School of the Griots, and the Negritude movement, the Freedom Culture of the 1970s emerged as a critique of these movements. (The proponents of Freedom Culture, at the outset, endeavored to distance themselves from what they deemed as literary movements.) They blamed the Indigenists, particularly the Noiriste wing of the movement, for the use of the Duvalier dynastic dictatorship. One leader of the movement exclaimed in response to this writer's mention of the intellectual productions of Indigenism: "it is Price-Mars who is responsible for all the things that the Duvalierists are doing today!" (personal communication, ca. 1976). Thus, we need to ask what was the philosophy of this movement, who its ideologues were, and how their contributions favored the advent of the present day Roots Music.

The "Freedom Culture," a movement that emerged outside of the country and purported to employ culture as a weapon to combat the social conditions of Haiti, the Duvalier dictatorship in particular, followed in the wake of the demise of the student movement inside the country in the 1960s. A student strike (in sympathy with that of the teachers and the labor unions), which shocked the foundation of the government and at the same time changed the political alignments within the student camp (due to the fact that many students rallied the Duvalier's camp, and subsequently became notorious *tonton makout* [Uncle Knapsack], a situation that caused the death, jailing, or exile of most of the opposition activists), was brutally repressed; however, the university students formed some cultural groups to promote the folklore of the nation, *Karako bleu* (blue denim), named after the denim shirt worn by the peasant, being one of the best known. Although the students managed to stage, at the French Institute and at other venues, some dramatic presentations, including a typical peasant wake, they could not last long. Many went into voluntary exile, and most were encouraged to go to the Congo, which following its independence crisis, needed Black role models for their youth and invited Haitians to work as teachers and in other public functions. The decade of the 1960s was the incubator for the politico-cultural movement of the next ten years.

The students and their elders (teachers, union members, political activists, and labor leaders) were not immune to the major events of the 1960s, notably the Civil Rights Movement in the United States and the Vietnam War. The progressive stance of the short reign of Pope John XXIII (1960–1963), in net contrast to the policies of his predecessor, favored the rise of a popular movement within the Catholic Church, no small power in a country like Haiti. The concatenation of these currents contributed to a spirit of rebellion, a yearning for justice, of course, with no obvious outlet for its venting, because the repressive structure of the country's government prevented all manifestations of discontent. As in previous times, the struggle shifted to the realm of the pen and paper, but this time it became clear that the Haitian peasant had to be enlisted and, since he could not read, other means had to be found to reach him. The use of folklore and the Creole language were then incontrovertible. The motivation of the Freedom Culture of the 1970s can be found in the realization of this important fact, arrived at after some painful setbacks.

The death of Papa Doc in April 1971 and the succession of his 19-year-old son, Jean-Claude Duvalier, was a turning point in the history of the opposition to the regime of oppression. Most of the opposition leaders were living in exile (those, of course, who had not been killed or jailed), either in Europe, Canada, or in the United States. Many of them were Marxists, in principle, who began to discover the important role of culture in the revolution. They reluctantly, in fact strategically, embraced the more palatable aspects of the nationalist thesis, which promotes culture and color as determinants in political conflicts. This meant that Vodun had to be included in the equation, if one were to reach the masses of Haiti, in order to lead them to the road of progress and civilization. Many Haitian Marxists, who upheld the Freedom Culture approach, considered Vodun as a "primitive religion" that, as in the case of all religions, may very well be a liability to the country's advancement; yet, the promoters of Freedom Culture embarked on a program that went to the very matrix of the Haitian nation. They decided to use the music of Vodun as the basis for their compositions, rightly realizing the fact that some 85 percent of the population had been using the songs of Vodun as their main learning vehicle. So, they changed the traditional words of the ritual chants and replaced them with their own. An example of this is "Ayiti demen" ("Haiti Tomorrow"); better known by its first line, "Le la libere, Ayiti va bel o"; composed in Boston (around 1974) by poet Koralen Jean-Claude Martineau of the group "Haiti Culturelle" on the tune of a well-known ritual song to St. Jacques Majeur/Ogou-Feray; and arranged by this writer, who was at that time the musical coordinator of the organization. They took a multimedia approach in the sense that they used dance, drama, music, and painting as means to impart their vision of the new Haiti. Were they fully aware of the centrality of the arts and culture in human life and, particularly, in a society like that of Haiti? Until further research is undertaken on this crucial issue, the answer to this question may have to wait. But, nonetheless, the Freedom Culture adherents struck a gold mine and, in this regard, were fundamentally different from any of the previous movements that purported to "liberate" the downtrodden.

Given the repressive nature of the Duvalier era, it is clear that the Freedom Culture could evolve only in exile. In fact, the movement rose in the major metropolitan centers of North America, like New York and Boston in the United States, and Montreal, where a substantial immigrant Haitian population lives. Three sister organizations, affiliated with the Patriotic Action Movement, the Marxist political wing of the opposition to Duvalier, were founded in the early 1970s to promote the role of culture in the struggle. In New York, it was Soley leve (The Sun Rises), in Boston Haiti Culturelle (Cultural Haiti), and in Montreal Vaccine (the bamboo trumpet featured in the Rara, the ritual spring festival of the Haitian countryside). It is important to mention that prior to and, to some extent, parallel to Soley leve, other politico-cultural groups like Tambou libéte and Kwi dor operated in New York, all of which contributed to advance the struggle for liberation through culture.

Two central points of their program were the use and promotion of Creole, as the language of the entire people, and the importance of the status of women, as an indispensable complement in the struggle. They also promoted a reinterpretation of

Haitian history, including the contribution of the peasantry, the backbone of the economy, the granary of the society, without whose labor few can eat. Songs such as "Haitian Women, Thank you, you are the Giver of Life," performed to the rhythm of the contredanse and *Peze kafe*, the story of peasant abuse on the way to sell goods to the market, played to a *Kongo siye* beat, illustrated these concerns of the movement.

The three sister organizations held joint events in each other's respective cities in order to raise funds and to reach out to the community. These events were multimedia and included vignettes, dances, songs, and jokes. They translated into Creole Jacques Roumain's *Masters of the Dew*, which they staged to great acclaim in Boston, Montreal, and New York. The translation of Roumain's celebrated novel was a significant act, when we remember that the novel had been translated into 17 languages (Michael Dash, Introduction to *Masters of the Dew* [Heinemann, 1978]), but never into Creole, the language of the people who are the subject of this literary masterpiece. To make sure that all Haitians had access to the message, they dramatized the story in a fully staged performance, which included a Vodun ceremony with a live rooster and battery of ritual drums.

They produced three recordings of Freedom Culture songs, with an assortment of didactic poems by Koralen, of which the best known are "Kamelo" and "Telson," whose central theme is the notion of popular justice in the face of Duvalierist oppression. The Freedom Culture activists circulated a number of cassette recordings throughout the Diaspora, as the expatriate community was called in those days, and some were even smuggled into Haiti. Thus, their contributions were felt far and wide and their message of hope and resistance penetrated circles deep inside the Haitian psyche. They vowed to change the artistic landscape to reflect their vision of a democratic Haiti: they criticized the popular musical culture whose song lyrics contained disparaging comments about women, like this one, "Fanm lan danse kou mabouya," (the woman dances like a fat lizard) referring to the lack of grace of a woman. The women within the movement played a prominent role. They did not spare their men, whom they often branded as "feudal" when the latter's behavior warranted it. Members of the Freedom Culture went to war against the trivialization of interpersonal relations found in so many of the popular songs of Haiti. They vowed to rehabilitate the woman, the center post, so to speak, of the Haitian family, and the peasant, the pillar of the economy, and predicted that the Freedom Culture ideology would soon come to replace the negative popular culture. They were visionaries who saw a day when the message of the Freedom Culture would become the "common denominator" of Haitian culture and dominate the musical productions of the 1980s and beyond.

While proponents of the Freedom Culture were working to raise consciousness about the plight of the Haitians, another group of young people, those who could not leave the country, used a different tactic, and they participated in Vodun rituals in the Haitian countryside. Their involvement in Vodun was a rather inoffensive gesture, since their actions were in accord with the Duvalier's program of "authenticity," which included the promotion of Vodun. Using the songs and rhythms of the *peristil*, the Vodun temple, which they went to learn at its source, they were about to begin a movement that would join force with their counterparts abroad to produce a revolution in the cultural arena of the 1980s and 1990s under the name of Rasin or Root Music.

The Rasin or Root Music

Although the Rasin Movement followed in the wake of the Freedom Culture, it is in the Vodun "lakou" or compounds that we must search for the roots of the present day progressive popular culture. The two movements did not evolve in total ignorance of each other, but the repressive atmosphere of the Duvalier regime made it difficult for their members to communicate, so the two currents developed independently, and we can understand that most people did not know about them until after the departure of Duvalier in 1986. Today's "mizik rasin" is grounded in the sociocultural/musical organization of Vodun, a system that requires some explanation, due to the tendency of too many to demonize this faith of the Haitian people.

Vodun, the religion of the Haitian people, retained from Africa, and practiced throughout the Americas under different names, is a comprehensive system of organization of the cumulative wisdom of humankind. In this sense, it is a science of universal knowledge that includes, among other things, social, political, economic, artistic, technological, and, more important, spiritual dimensions with roots deep in the history of the universe. Vodun teaches that all information and resources must be put to the service of the community for all to enjoy, regardless of their economic status. Central in the philosophy of Vodun is the concept of justice and equity and the obligation to serve the needs of the communal whole. The teaching of Vodun is transmitted through the oral tradition by means of its songs and dances, accompanied by a battery of three drums perfectly harmonized in a balanced symphony of movement, sound, and sight, and orchestrated by the presiding *lwa*, or the spiritual entities that the community serves. In Haiti, this accumulated knowledge has been preserved in the "lakou," or the agricultural compounds, the foundation on which the social organization of the countryside rests. And it is there that the proponents of today's Root Music went to search for their heritage.

According to Wilfrid Lavaud a.k.a. Ti Do, one of the leading members of the group Foulah (personal communication 1993), the proponents of the movement traveled to the provinces every weekend, often at great risk, for they were always under the watchful eyes of the *tonton makout* (Duvalier's militia), to study the music of Vodun in its natural milieu. They learned to play the rhythms of the Rada, Kongo/Petwo, and other rites in "lakou" like Souvenance and Soucrie in the Artibonite Valley, and others like Lakou Badjo, Lakou Malik, and Lakou Lan Kanpech. They also took music lessons, and they began to transfer the traditional beats to their guitars and other modern instruments. In their compositions, they introduced the ritual drums and bamboo trumpets, known as "vaksin," much as their predecessors of the Freedom Culture did. While they tried to respect the integrity of the music, they also experimented, at times quite successfully. They formed musical groups that they named Foulah, the name of the mouthful of rum the priests spray ritually over the participants at ceremonies, alternatively, the kerchief ritual drummers and other Vodun adherents wear around their neck; "Boukman Eksperyans," the name of the Vodun at the Bwa Kayiman congress in 1791, an event that changed the history of St.-Domingue, or again, "Boukan Ginen," meaning the flame of Africa. These groups, formed after the departure of Duvalier, emerged on the scene during the late 1980s, and during the coup d'etat (1991–1994) the group RAM, directed by Richard A. Morse, emerged to continue the consciousness-raising work in Haiti, while the others mentioned above were forced to go into exile. Many of their songs use the double entendre technique, a procedure well known in Vodun to avoid the wrath of the military dictators. A good example is the song "Embargo" that the military liked, but it meant much more than that, because one could assume that it referred to the military's grip over Haiti. Another song of RAM, "Ibo Lele," which employed the *ibo* rhythm as its constructive material, was used in Jonathan Demme's movie *Philadelphia* (1994), giving exposure to the new music of Haiti. The Root Music groups achieved international fame with their recordings and concert tours and carved out a niche for the new Haitian music in the world market.

The best known among these musical formations is Boukman Eksperyans, founded in 1989. It should be said that some of its musicians left to form the no less famous Boukan Ginen (1993). Thus, in a sense, the two groups share a number of similarities, both in musical approach and philosophy. Boukman Eksperyans's song "Nou Pap Sa Bliye," an ode to the spirit of Bwa Kayiman and dedicated to Lakou Souvenance, and Boukan Ginen's "Nati Kongo," a song of remembrance, articulate the message of unity, concern for Haiti's impoverished peasants, and intolerance for political corruption and neglect, which struck a chord with a community identity and direction after a long period of political searching (Averill, 1990). The young promoters of Root Music (Ti Do, Bonga, Richard A. Morse, the Beaubrun brothers and sisters, Aboudja, Kompe Filo, Eddy François, Pierre-Rigaud Chery, Jimmy Jn-Felix, Carline, Manina, and Manze, to mention but a few) differ markedly from their Freedom Culture forebears (although, veterans like Nikol Levy, the musical director of Soley Leve, and Sò Ann perform with Foulah regularly); Guy Gilbert of Vaccine and composer of the song "Sezon Papayon," which he wrote in memory of his brother Delano, whom he thought had died in jail, is now the Haitian Consul in Montreal; and Jean-Claude Martineau of Haiti Culturelle in Boston served as President Aristide's spokesperson, in the sense they have come to terms with the all encompassing philosophy of Vodun. The Freedom Culture adherents were less concerned with the content of Haitian culture because many of them, in their Marxist or petit bourgeois orientation, saw the culture as being backward, and therefore something to be transcended or transformed. How convincing they were in their commitment to equality is open to question. Thus, it is not surprising that a few of them jumped ship and joined the Duvalier regime in its waning days, and some even went on to serve the various military regimes of Avril and Cedras. The present day Root Music protagonists, having done their homework, came rightly to see Vodun as an expression of African culture, "a mother-lode that must be mined" (Bernice Reagon Johnson, lecture at Tufts University, ca. 1983). In this sense, they were poised to harness the spiritual energy and goodwill embodied in the age of consciousness spawned by the Civil Rights Movement and its aftermath in the United States, the struggle for African liberation on the continents and the rise of the wretched of the earth.

The music of Rasin emerged at a time when a good part of the world seemed ready to embrace progressive humanism, spawned by the world music movement, a trend I call *Humanocentrism* a philosophical approach that goes beyond the oppositional concepts of Afrocentrism and Eurocentrism to assert the centrality of the person, the

Homo sapiens, regardless of ethnicity, class, gender, or lifestyles, in all human and artistic endeavors. In this sense, the Rasin Movement of the 1990s benefited from this favorable climate in which the reggae of Jamaica; the "Bloco Afro" of Brazil (Olodum, Ara-Ketu, and others); and the "Mezcla" of Cuba (Lázaro Ros and the Muniquitos), themselves preceded by the "New Song" of Venezuela and the Dominican Republic, were all evolving. Central in all these movements of cultural affirmation and liberation is Haiti, whose gesture in Bwa Kayiman in 1791 and Gonaives, the site of the Declaration of Independence in 1804, struck the imagination of the world. Thus the themes treated in the songs of the Root Music Movement meet not only the aspirations of the Haitian people for dignity, peace, and justice, but they also strike a resounding chord throughout the world progressive community.

Conclusion

Vodun, as a comprehensive system of universal knowledge, with its emphasis on the artistic and spiritual components of life, summarizes the very soul of the Haitian people; in effect, it is in the arts of the popular culture, the music, the dance, the songs, the paintings, the richly decorated flags, and other ritual objects, now profusely produced for the outside market, that the religion's contribution is most evident. The arcs of the popular culture, the folklore of Haiti, grounded in the African past and eloquently expressed in the songs, are a tribute to the people's aspiration to freedom. Traditionally, Vodun songs are poems of affirmation, whose themes cover the whole spectrum from the didactic and the utilitarian to the revolutionary, from the contemplative to the sublime, and from the transcendent to the recreational. And, it is not surprising that the song of freedom, the Vodun song, would inspire a popular movement of conscientization whose career is treated in this chapter. The coming together of the Freedom Culture and Rasin or Root Music movements, resulting from a long awaited apprehension of the true heritage of their ancestors from Africa, as expressed in the literary concepts of Indigenism, Negritude, the Harlem Renaissance in the United States, and the artistic productions spawning from the 1946 bourgeois social revolution, is breaking new ground on the road to qualitative social revolution, the progressive humanization of world culture, or in one word, Humanocentrism. The fruits of these artistic and literary currents are already felt in the works of an illustrious crop of individual musicians such as Farah Juste, Myriam Dorismé, SÒ Ann, Barbara Guillaume, Carol Maroulé, Manno Charlemagne, Jn-Claude Eugène, and groups such as Antidote, Phantom, Rara Machine, Missiles, and Zin, which make it a point to promote and disseminate the message of hope, freedom, justice, equality, peace, progress, and unity. The prediction of the proponents of Freedom Culture in the 1970s that their philosophy would become the common denominator for Haiti's music is about to be realized, for there are few popular music groups today that do not include in their repertoire a composition or two based on the conceptions of the song of freedom. The result of all of this is that the Haitian population, regardless of its high illiteracy rate, is among the most politicized in the world. And now, we only need to put it all in a meaningful educational form, a package, one may be tempted to say, that mines this mother lode systematically in a way that serves the material and spiritual needs of the Haitian as well as that of his sisters and brothers throughout the world.

Selective Bibliography

*Averill, Gage, 1990. *One Day for the Hunter, One Day for the Prey: Popular Music and Power in Haiti.* Chicago: University of Chicago Press.
Césaire, Aimé. 1971. *Discourse on Colonialism.* New York: Monthly Review Press.
DuBois, W. E. B. 1969 [1903]. *Souls of Black Folks.* New York: New American Library.
Fanon, Frantz. 1959. *The Wretched of the Earth.* New York: Grove Press.
Fleurant, Gerdès. 1989. Caste, Class, and Status Quo in Haiti. In *Ethnic Conflict and Power: A Cross-National Perspective*, ed. Gelfand and Lee, 178–193. New York: John Wiley.
Freire, Paolo. 1972. *Pedagogy of the Oppressed.* New York: Herder and Herder.
Hunt, Alfred N. 1988. *Haiti's Influence on Antebellum America: Slumbering Volcano in the Caribbean.* Baton Rouge: Louisiana State University Press.
James, C. L. R. 1963. *The Black Jacobins: Toussaint L'Ouverture and the San Domingo Revolution.* New York: Vintage Books.
Leyburn, James G. 1966. *The Haitian People.* New Haven, CT: Yale University Press.
Manuel, Peter. 1988. *Popular Musics of the Non-Western World.* New York: Oxford University Press.
Nkrumah, Kwame. 1970. *Consciencism: Philosophy and Ideology of DeColonization and Development with Particular Reference to the African Revolution.* New York: Monthly Review Press.
Price-Mars, Jean. 1928. *Ainsi Parla l'Oncle.* Paris: Impr. de Compiege.
Roumain, Jacques. 1978. *Masters of the Dew.* Portsmouth, NH: Heinemann.
Williams Yarborough, Lavinia. 1964. *Haiti-Dance.* Frankfurt am Main: Bronners Drückerei.

Selected Discography

Ayibobo. Cine Disc, Port-au-Prince, 1993.
Boukman Eksperyans. *Vodou Adjae.* Mango Records, NY 1991.
Boukan Ginen. *Joun a Rive.* Cine Disc, Port-au-Prince, 1993.
Carole Maroulé. Fleetwood, CDM 1277 (ca. 1983).
Charlie Mingus. *The Best of Charles Mingus.* Atlantic Records, SDI 555, 1970. RAM.
Sò Ann. *Gad. our Zetoik,* Marc Records. NY JDJ 1839–1.
Soley Leve. *Chansons et Poiries Patriotiques.* Haiti, Soley SOS 001 1973, vol. I; *Peyi an Mouin, SL-1947A. Vol. II; Ayiti Demain Pete Kafe, 1976, Vol. III.*

Chapter 5

Yon Moso Twal Nan Bwa (A Piece of Cloth on Wood): The Drapo Vodou in Myths of Origin

Anna Wexler

Tout sa m ap di ou, se yon istorik li ye. Gen pou ale nan kòmansman pou rive sou drapo a . . . Manbo Ayizan Potko se gwo rèn Vodou—se li vin parèt avèk Vodou. Tout Lwa yo rete la avèk li men yo potko eklere. Yo pa konne kilès yo ye, kisa pou yo fè. Sa se lontan lontan. Yo tout rete kay Manbo Ayizan paske se li ki pi gwo, semanman tout oungan se li ki responsab pou gade moun nan djevo e ki konnen tout règleman pou sèvi mistè. L ap kouche moun, l ap trète moun, li gen sa k pou pile fèy, sa k pou okipe moun. Kounye a Papa Loko ki responsab ason konmanse revolte. Papa Danbala, li kite kay Manbo Ayizan tou e li rete nan palmist kote li vin tounen koulèv. Agwe Tawoyo tounen pwason, li konmande lamè. Ezili Freda se menm pawof la. Li vin abite a lame. Se yon dam ki renmen anpil bijou, pafen. Kouzen li menm leu li te rete kay Manbo Ayizan, li te konn netwaye tab. Zaka, li te konn fè jaden. Li te ale tou, li fè kilti pa li, li travay latè. Chak Lwa ki revolte vin responsab yon bagay, li senbolize yon bagay. Yo defile kay Manbo Ayizan youn pa youn. Kounye a Papa Sobo di bon, m pral fè drapo. Li pran on moso bwa, li mete yon ti twal ladan, li di bon, l ap gen drapo pou siyale tout Lwa yo. Kounye a se li menm ki responsab drapo, se li ki mèt drapo. Li fè yon drapo blan, yon drapo wouj . . . lape, viktwa. Gen yon jenn ti gason ki rete avèk Manbo Ayizan tou. Sobo soti avèk ti jenn nomn nan, kounye a li di/li sè Laplas la kay ou yé. Li soti avèk ti nèg la Laplas ia. Kounye a Papa Sobo pran drapo, ak li e pi chak pran plas yo. Sobo soti avèk Bade tou, ti frè li, yo tout soti ansanm ak drapo/a. Drapo a vin rele Sobo e Bade.

Everything I'm going to tell you, it's a history. You must go to the beginning to get to the flags . . . Manbo Ayizan is the great queen of Vodou—she was the first to present Vodou. All the *lwa* [spirits] lived with her but they were not yet enlightened. They didn't know who they were, what they were supposed to do. It was long long ago. They all lived in Manbo Ayizan's house because she was the most powerful, the mother of all the *oungan* [priests] who was responsible for watching over everyone in the initiation chamber and knew all the rules for serving the spirits. She initiated people, she treated people, she had what was needed to crush leaves, to take care of people. Then Papa Loko who is responsible for the *ason* [rattle] began to revolt. Papa Danbala, he also left Manbo Ayizan's house and stayed in a palm tree where he turned into a snake. Agwe Tawoyo turned into a fish, he commands the sea. Ezili Freda was in the same group. She came to live in the sea, she is a woman who loves jewelry, perfume. Kouzen himself when he lived in Manbo Ayizan's house he used to clean tables; Zaka he used to work in the fields. He left too, he tends his own crops now, he works the earth. Each *lwa* who revolted became responsible for something, symbolized something. They filed out of Manbo Ayizan's house one by one. Then Papa Sobo said, "Good, I am going to make a flag." He took a piece of wood, he put a little cloth on it, he said, "Good, there is going to be a flag for signaling all the *Lwa*." Now he's the one responsible for the flag, the master of the flag. He made one white flag, one red flag . . . peace, victory . . . There was a young man who also lived with Manbo Ayizan. Sobo left with the young man and then told him, "You are the *laplas* of the house now." He left with the young person, the *laplas*, and then Papa Sobo took the flag, each took their place. He left with Bade too, his younger brother, they all left together with the flag. The flag came to be called Sobo and Bade.[1]

Shortly after our first meeting in Miami in 1991, the Vodou flagmaker Clotaire Bazile introduced me to his spiritual history of the form with these words: "Gen pou ale nan fon pou rive sou afe drapo a" (One must go to the bottom of things to get to the history of flags). In 1993, toward the end of my doctoral research that focused on his ritual artistry, he finished his account of the origin of the flag as transcribed above. With his characteristically acute sense of timing, he chose to complete it when I could hear it as more than an exotic narrative; when I could see Sobo's bit of cloth wave as the original signal for the *Lwa*, at the threshold of form, of visibility, when the spirits first emerged into particularized being, for us and then, repeatedly, through us. In the intervening years, I had encountered *drapo* (flags) in multiple ritual contexts and in diverse narratives of their sources. Through this deepening but circuitous inquiry, I eventually returned *nan fon*, and finally I prepared to receive the concluding moments of the flag's genesis from Clotaire. There *are* other origin myths. Much of the recent scholarly literature on the *drapo* Vodou is focused on the question of sources contributing to the visual form of the flag and the ritual protocol that the *kò drapo* (flag corps), including the *pòt drapo* (flagbearers) and the *laplas* (ritual assistant and swordbearer), enact. Taken as a whole, these accounts, drawing variously on multiple African and European military and religious influences, constitute a prismatic narrative of creolization, of the simultaneous transformation of sources, in which one face of the elusive and multivalent form cannot be exposed without revealing another. These accounts too may begin with "yon moso twal nan bwa" but may be held aloft in other struggles.

In *The Four Moments of the Sun* (Thompson and Cornet 1981) Robert Farris Thompson, the first to map African genealogies, linked the S-shaped cross-guards sometimes found near the pointed tips of the staves of Haitian ritual flags to a Kongo gesture: the "crossroads pose—right hand up to heaven, left hand parallel to the horizon line" (171). This gesture, found on Kongo *niombo* or reliquary figures and on the handguards of royal swords used for executions in ancient Kongo, is said to mark "the boundaries between two worlds" (172) and to appear on the Haitian ritual flag that signals the spirits and marks the limits of ritual space in a possible creolized restatement of this theme. *Nkusa minpa*, the Kongo practice of unfurling and waving cloth to open the entrance to the spiritual world, is also described as an antecedant to the ritually mediating role of the liminal *drapo* Vodou. Thompson reemphasizes this link based on the sacred activity of cloth in both cultural contexts in *Flash of the Spirit*, where his major discussion of Vodou flags occurs under the heading of "Kongo Influences on Haitian Sacred Art" (1983:179). The glitter of beaded figures of the *lwa* on the *drapo* is said to "suggest the presence of spirit" (186) just as bits of mirror and other reflective materials on Kongo *minkisi* ("power figure") are described as indicating "the flash and arrest of the spirit" (118). However, the S-shaped cross-guard on the flag stave has moved from Kongo gestural resonances to Dahomean military/religious associations in the *gubasa* or sword of the martial deity Gu. Thompson suggests that the S on the Vodou flagstaff reverses that of the hand-guard on the saber held by St. James, identified as Sen Jak, the chief of the Haitian Ogou, in the Catholic chromolithograph, thus linking swords and flags in a creolized extension of the spiritual assertion associated with the Dahomean *gubasa* and the Fon warrior god who wields it.

In the more recent *Face of the Gods* (1993), which contains his major writings on the sacred flag in the African-inspired religions of the Atlantic world, Thompson expands on his sense of the Kongo contribution to this form as it developed in Cuban Regla de Mayombe altars, Brazilian tree/flag altars for the spirit Tempo, and Suriname maroon shrines. Though not discussed at any length, the spirit signaling, danced Vodou *drapo* is linked to the latter, more stationary flag/altar expressions and hence to Thompson's interpretations of their Kongo antecedents:

> Flags are important ritual presences in Haiti as in Suriname and Cuba . . . Their function seems cognate with that of the *banderas* mystically moving *prendas* in Cuba, of the flags that announce the presence of Tempo in Brazil, and of *faaka pau* and *faaka tiki* standing to mark the spots where spirit can be praised among the maroons of Suriname. In Haiti, flags come off the T-form hoists they ride in Suriname, or off the staffs of Tempo or the closet walls of *minkisi* in the Cuban Bronx and join the faithful in the dance. (1993:294)

Discussing Kongo tombs as altars, Thompson explains that *mpeeve*, the Ki-Kongo word for flag, also means " 'wind, breeze, spirit' " (22) and that the flags placed on the ends of tombs indicate the movement of the dead as "spirit in the wind" (22). "Spirit, moving in cloth," he writes further, "pinpoints in space a Kongo altar" (72–73) and is also linked to the *nganga* or priest trembling in ecstasy. This association of the flag with the activation and movement of invisible spiritual forces is basic to his reading of African Atlantic flag/altar forms as Kongo-inspired. For example, in his discussion of altars for the spirit Sarabanda in the Kongo-influenced Cuban tradition of Regla de Mayombe, he describes a red *bandera* or flag on the wall behind a *prenda* (sacred cauldron) as deepening Kongo resonances and "allowing it (the *prenda*-author) to move through time and space" (68). Felipe Garcia Villamil, a Cuban *mayombero* living in the Bronx, is quoted as saying " '*Prenda no camino si no hay bandera*' (Without a flag, the *prenda* won't move)" (qtd. in Thompson, 1993:68).

Kongo flag traditions, in addition to flag-surmounted Muslim marabout graves and Mande/Akan tree altars, are also invoked in Thompson's writing on Samaaka ancestral flag altars, *faaka pau* or " 'staff with a flag' " (1993:129). He associates this form—flags on trees or suspended from tall poles on a platform in the air with a ladder leading up to it—with *nsunga* or Kongo banners attached to long poles held during funeral processionals and used to signal the dead. It is interesting to note that similar forms have been documented (though rarely) in connection with Haitian Vodou. For example, one of the photos accompanying the New Library of America edition of Zora Neal Hurston's *Tell My Horse* (1995), based on her research in Haiti in 1936–1937, is labeled "Offering to Congo Savanne," a group of *Lwa* in the Kongo pantheon. Several flags, most of them white, are suspended from what seems to be rope twisted up the trunk of a tree. Beside it there appears to be a ladder reaching up to the level of the flags. In *En Haïti* by the French planter Eugène Aubin, published in 1910, photographs documenting the beginning of a procession to celebrate a service for the *Lwa* Agwe in Grand Saline include plain cloth banners (a few in white, one vertical swallow-tailed flag in dark translucent material) suspended from long stationary poles outside the *ounfò* of *ougan* Ti-Plaisir. In the foreground of one photograph, a *ounsi* (temple) carries a flag—a piece of dark translucent cloth on a short stick—a more familiar, basic version of the mobile Vodou flag.

In Patrick Polk's essay, "Sacred Banners and the Divine Cavalry Charge," in the exhibit catalogue *Sacred Arts of Haitian Vodou* (Cosentino 1995), the S-shaped crossguards found on the staves of some *drapo* Vodou are situated explicitly in Haitian colonial and postcolonial military history and linked to similar features found on European-style military standards and swords. Emphasizing the impact of European flags on indigenous forms in sub-Saharan Africa, including Fante, Dahomean, and Kongo traditions, Polk goes on to explore the influence of specific kinds of French colonial flags (especially Napoleonic regimental colors and Royalist and Republican banners)—their adoption by insurgent troops and the postrevolutionary Haitian military, their status as authorizing and sacralizing objects, and the ceremonial protocol associated with them—on the form and ritual choreography of the *drapo* Vodou.

As Polk (1995) and Yonker (1991) have pointed out, elements of their ritual deployment, such as the prescribed movements of flag escorts, pairing of flags and swords, crossing flags over the heads of honored participants, extending the tips of the flags and the hilt of the sword to be kissed and dipping in salutation, as well as the circular runs around the *poto mitan* (the center post of the *peristil* or Vodou temple), quite obviously reflect the impact of the military reviews and salutes carried out by the colonial forces. Although, as Polk documents, there is evidence that flags belonging to Vodou societies were in use by the mid-1800s, to my knowledge the details of their actual deployment in ceremonies during this period have not been systematically researched. The earliest references I have located to flags being used to lead processions for the *Lwa* and to salute members of the ritual hierarchy and honored guests are those contained in the writings of Eugène Aubin, mentioned above, which are based on observations made at the turn of the century. Described by Métraux (1972) as an accurate observer of Vodou, he documents ceremonies in LaPlaine and Carrefour in which the *ounsi* unfurled *drapo* and, accompanied by the *laplas*, saluted dignitaries. In the latter locale, they led a *manbo* as she made libations under the trees sacred to the *Lwa* around the courtyard. When visitors entered following this procession, the *ounsi* placed the tips of the flag staves on their shoulders. In Aubin's photo of a procession for Agwe in Grand Saline, also mentioned above, the *ounsi* in the foreground holding a flag of transparent, dark cloth appears to be waiting to lead a group of *sèvitè* gathering in the background.

In Polk's account, the sacred associations of the Vodou flag draw on the "aura of divine power" (1995:336) with which the French embued their flags as embodiments of colonial authority in St.-Domingue and with which they sacralized their regiments. As he points out, military flags were customarily blessed before being carried by the regiments and were kept in churches when not being used. Included in *Haïti et ses Peintres* by the Haitian art historian Michel Philippe Lerebours (1989) are reproductions of two paintings that (unintentionally) document the complementarity of the sacralized military/national and Vodou flag that Polk suggests. One work entitled "La Bénédiction des Drapeaux" (undated) attributed to Louis Rigaud, who became recognized as a painter in the late nineteenth century, depicts a pair of flags, including the national flag of Haiti, being presented by Haitian troops for consecration to a French Catholic priest. In the other, more contemporary work, "Bénédiction des Drapeaux par la Maitresse de l'Eau" (undated), by the painter Gérard Valcin (1927–1988), the sacralizing agent is a divinity, the mermaid *Lwa* Lasyrenn. Although it is difficult to make out the details of this painting in the reproduction, one can detect a *ounsi* standing in the water and holding out a flag (or perhaps a pair of flags) for blessing by the female water spirit who sits on a ceremonial bark in a stream in front of a waterfall. A number of *ounsi* and other participants also stand in the water in an attitude of prayer.

A quintessential expression of the fusion of religious and political power in Haitian history, the Vodou flag and its ritual, in Polk's reading, ceremonially reenact the transformation of European military symbols, technology, and discipline into the revolutionary weapons of enslaved Africans and their descendants and re-create their victory. Connections between revolutionary heroes, the first flag of the nation, and the Vodou flag are evident in the iconographic emblems and activities of the Ogou, especially Ogou Feray and Sen Jak, whose image accompanied by his flag from the chromolithograph of St. James the Major frequently appears on *drapo* Vodou. As Desmangles (1992) suggests, these *Lwa* are associated with the Haitian generals and soldiers who defeated the French as well as the guerilla forces, the Cacos, who fought the Marines during the American Occupation of Haiti (1914–1934). Karen Brown writes that "[t]he military-political complex has provided the primary niche for Ogou in Haiti" (1989:71). A painting by Senèque Obin entitled "J. J. Dessalines" (c. 1960)—the Haitian Revolutionary general and first Haitian head of state—makes the links between Ogou, Dessalines, and the first flag of Haiti explicit. As Stebich (1978), in whose book *Haitian Art* the painting is reproduced, explains, the red color of his uniform as well as his pose in the painting evoke Ogou. The red and blue flag he holds is the flag of the new republic he created in 1803 by tearing the white stripe off the French tricolor, enacting the defeat and expulsion of the colonists (Trouillot 1990). Legendary accounts of this moment abound. According to Odette Mennesson-Rigaud (1958), people still discuss whether the flag was created by Dessalines under a *repozwa* (tree housing a spirit) near a *ounfò* in Merote, in blue and red for the Ogou he served, or in a meeting house in Arcahaie. In her book, *Haiti, History, and the Gods*, Joan Dayan (1995) writes that people in Léogane told her that Dessalines was possessed by Ogou when he cut the white stripe off the French flag. In a ceremony for Ogou Feray described by Milo Marcelin (1950) in *Mythlogie Vodou* the *Lwa* is presented with two flags, the flag of the Vodou *sosyete* and one in blue and red, the colors of the nation, and as Marcelin explains, of the spirit.

In the mythic resonances between the first flag of the nation and the Vodou flag, the narrative of transformed sources surrounding the ritual *drapo* restates the unity of spiritual and political rebellion integral to the birth of the first Black republic in the New World through the legendary origins of its most emblematic military/religious form. Whatever the truth status of Bois Caïman, the Vodou ceremony and political rallying event of 1791 believed to have launched the Haitian Revolution (see Mintz and Trouillot 1995), it is clear that the religion was a vital tributary to the slaves' resistance to and eventual military defeat of the colonists.

Not to be neglected in this facet of the prismatic origin narrative of the *drapo* Vodou is the monumental treatise on the occult political/religious significations of the Haitian national flag by Dr. Arthur Holly a.k.a. Her-Ra-Ma-El, self-described as "Esoteriste Haïtien" and, according to Magdaline Shannon (1983), a scholar respected by his contemporaries in the Haitian elite for his knowledge of the esoteric dimensions of Vodou. In a letter to his brother, then president of the Ligue du Drapeau, which prefaces this volume, Holly states his intent to reveal the "valeur occulte du Drapeau haïtien" (the occult meaning of the Haitian Flag) (1928:14). Entitled *Dra-Po, Etude Esotérique de Égrégore Africain, Traditionnel, Social, et National d'Haiti*, this book was published in Port-au-Prince in 1928, the same year as *Ainsi parla l'oncle* by Jean Price-Mars. The latter, a groundbreaking work of Haitian Indigénisme (the valorization of Haiti's African heritage), sought to valorize Vodou as a source of national pride and unity, partly by linking it to African religious traditions, and to counter a particularly virulent wave of racism defamation of the popular religion spurred by the American Occupation (Murphy, 1990). According to Milo Rigaud (1953), Holly's oeuvre represented a continuation of the efforts of Price-Mars and his colleagues. However, his construction of African sources is an occult project, tracing hidden meanings within the symbolic codes of Vodou to initiatory Ethiopian and Egyptian traditions, including ancient Chaldean astronomy, in an effort to dignify the popular religion and combat racist images of its practices as cannibalistic and obscene.

In *Dra-Po*, occult readings of the Haitian flag, its predecessors and its variant post-independence forms assertively fuse esoteric African genealogies of Vodou and theocratic versions of Haitian history with the cultural nationalism of the period. The flag, "un fragment de l'esprit, de l'âme, du corps de la Nation" (a fragment of the spirit, of the soul, of the body of the Nation) (1928:19), does not symbolize as much as enact the spiritual/political evolution of the Nation. According to Holly, even the colors chosen for the flags of the colonial troops were directly inspired by "des Invisibles qui sont les Protecteurs naturels de la race africaine" (the Invisibles who are the natural Protectors of the african race) (23) who were "mises en activité" (put into action) (24) by the colors representing them and thus able to "contrarier" (thwart) and "paralyser" (paralyze) the enemy from behind the lines. The red and blue bicolor created by Dessalines is said to have invoked the occult powers of Mars and Jupiter or "Ho-Gou phérail" (Ogou Feray), who, like other *Lwa*, represents a cosmic law or force augmented by advances in science and other achievements of human reason. Holly proposes to convene his readers under "l'ombre du DRA-PO" (the shade of the flag) (36) in order to penetrate into the realm of national heroes and ancestors, reclaim Vodou from its detractors, and revive the "l'âme nationale" (national soul) from the lethargy induced by the ethical and political disarray of the period—a clear reference to the humiliating impact of the American Occupation. On the revolutionary battlefields, around the *peristil*, and in the occult configurations of the text, the flag invokes and activates the spirits that unify and sustain the nation and its citizens in the face of continuous threats to physical and moral survival. Whether one reads from origin myths of the national flag to the Vodou flag, or the reverse, the distance between them effectively disappears in Holly's treatise.

For Louis Maximilien, whose classic work *Le Vodou Haitien* (1945) is dedicated to Price-Mars and extends the tradition of sympathetic, ethnographically informed portrayal of the popular religion, Vodou is "un culte du drapeau" (a cult of the flag) (144). By this he means that the patriotic character of the religion, embodied in the consecrated Vodou flag, is based on respect for and defense of the land considered to be the sacred ground of the ancestral *Lwa* and not just the object of secular claims. In his view, the military dimension of the ceremonies also refers to the vital support of the *Lwa* during battle or at other times of mortal danger. The spirits who may "chantent, dansent et crient, quand ils bambochent" (sing, dance and cry when they carouse) (145) can become instantly concentrated and aggressive when called upon to aid a vodouist in crisis. It is interesting to note that the Haitian novelist Marie Chauvet (1960), who cites Maximilien as an ethnographic source for descriptions of Vodou rituals in her novel *Fond de Nègres* (see chapter 2 of her book), evokes associations between the flag of Ogou and the reclaiming of peasant lands in the dreams of Papa Beauville. The red flag becomes an agent of cleansing and refertilizing the earth whose sale and misuse (including the chopping down of sacred trees) has angered the ancestral *Lwa*.

The contributions of French Catholicism are the least explored among the major tributaries constituting the multifaceted origin narrative of the form and ritual functions of the Vodou *drapo*. European military sources for the Haitian ritual *drapo*, as Polk (1995) indicates, include Catholic rites of sacralization by which banners were blessed by priests before being carried by the colonial regiments. The consecration of Vodou flags, and all other objects intended for ritual use, referred to as baptism or *batèm*, formally resembles the Catholic practice with a *prêt savann* or bush priest often officiating, described by Desmangles as a "symbol" (1992:89), but not an official representative of the Catholic Church, who assists *oungan* and *manbo* (especially in urban areas) with aspects of Vodou ritual requiring Catholic liturgy. Prayers are recited, holy water is sprinkled on the object, and a *non vanyan* or ritual name chosen by attending godparents is given to the object (Métraux 1972).

In *The Faces of the Gods*, the Haitian scholar Leslie Desmangles argues that baptism in Vodou is fundamentally shaped by West African rather than Catholic conceptions because it is used "to denote a means whereby divine power is instilled in a person, an edifice, or an object" (1992:90). Lionel Hogu[2] has also described the common Catholic practice of baptizing objects and dwellings in Haiti in these terms, as effecting the infusion of the protective and generative power of God into a house, for example, which may be baptized by a priest at the request of the prospective inhabitants before they move in. Catholic baptismal rites in Haiti, like other sacraments of the Church, evidence the long-term transformative impact of the religious cultures of enslaved Africans and their descendants who adapted them to their own needs (see Dayan 1995), whether in the Church or the *peristil*, a New World manifestation of a process dating from the fifteenth century when Portuguese missions were first established in West and Central Africa (Sanneh 1983; Thornton 1992). Vodou rites of baptism, then, do more than borrow superficial gestures from a Catholic practice that has itself been deeply shaped by African religions and their transmutations in Haiti.

Even in its most abbreviated form, the baptism of *drapo* intended for ritual use profoundly alters their status as objects. Clotaire once referred to Vodou flags made for sale, including the most elaborate and expensive versions, as "nothing" compared to those that have been consecrated. According to the flagmaker Edgar Jean-Louis, unbaptized flags "pa gen nanm" (don't have soul, energy), unlike consecrated *drapo* that can acquire even "*plis nanm*" (more soul) "*fòs*" (force), and "*kouraj*" (courage) by participating in ceremonies. Even when faded, torn, and bloodstained, the sacralized flag, especially those that have been involved in many rituals, are more highly valued by Jean-Louis than the most glorious unbaptized commercial flag.[3]

Sacralizing the flags may involve more elaborate and intense rituals like those undergone during the second level of Vodou initiation known as *kanzo* (initiation), described by Brown as a "rite of fire designed to transform suffering into power" (1991:351) in which initiates must handle dumplings pulled from boiling pots called *zen*. Jean-Louis used the terms baptize and *kanzo* interchangeably for the process of consecrating the flags. In the version of this ceremony he described, a mixture of oil and *fèy* (leaves) are prepared in a pot. As a mixture burns and the flames leap up, the flag is passed through them. On the final day of the three day ritual, lasting from Friday until Sunday, a dance is held to celebrate and formally present the flag to the *Lwa* for whom it is intended.[4] For the flagmakers Monique and Pollone Colin, to *kanzo* a flag is to give it "*plis fòs*" (more force) than the *prêt savann* can confer through baptism. In their version, the flag is sequestered in the *djevo* (initiation chamber) for up to eight days and made to *kouche* or lie down, like human candidates, on the sign of the *Lwa* to whom it is dedicated. There it rests on a special bed surrounded by a white curtain until "ou leve li" (you get it up) and then "ou resevwa drap a" (you receive the flag) with a meal consisting of chicken killed for the occasion, rice, plantains, yams, spaghetti, and kola that is placed beside the flag for the *Lwa* within it to consume. After the meal, the flag is put to bed again and brought out of the *djevo* the following day ready to assume its public ritual functions.[5] Sacralizing flags through *kanzo* rites does resemble those means used by the Fon and their neighbors in Benin and Togo to activate *bocio* sculptures for spiritual work (including heat, sacrifice, food, and invocation of deities) (Blier 1995) more than the Catholic ceremony of baptism.

The most important link between Catholic tradition and the ritual form and functions of the Vodou *drapo* is explored in LeGrace Benson's (1996, 1997) work on Catholic processional banners. In 1989 she observed striking similarities between the use of these banners in a Catholic ceremony known as a Pardon in a pilgrimage parish in Brittany and the ritual protocol of the Vodou flag. Brought out of a sanctuary in Locronon, where Pardons occur at six-year intervals, banners and crucifixes saluted the four cardinal points in the square in front of the church and were then kissed at each situation, ritual gestures commonly enacted in the ceremonial presentation of the Vodou *drapo*. As in Vodou services, these aspects of flag protocol have a distinct military valence. Later the banners were carried out into the town and surrounding fields in a sacralizing procession that stopped for prayers and hymns at each station. Relating this circuit to the ritual perambulations of Vodou flags outside the temples (and to military parades) LeGrace Benson writes beautifully,

> The intent and content of the procession may have been quite different from the processions of vodou banners into the surroundings of the *houmfo* [temple] during special services, but, like the parades from the President's or General's Te Deum in the cathedral out to the Champs de Mars, the poetry is the same: the drapo/drapeau proclaim the dominion of those who carry them and of their holy figure." (1996:7)

According to Benson, such services and processionals may have been transplanted to Haiti by Breton priests, who constituted about 70 percent of the clergy active in Haiti between 1804 and the 1860 Concordat with the Vatican, particularly in ceremonies dedicated to St. Anne, a popular focus of veneration in Brittany and Haiti. In addition to elements of shared ritual protocol, Benson notes significant visual crossovers between Catholic banners and Vodou *drapo*—the image of a saint in the center as well as the Sacred Heart of Mary and of Jesus transmuted into the heart *vèvè* used on flags to represent the Ezilis, deities of love.

To search for the sacred Vodou flag in its entangled histories was to arrive *nan fon* through violently disjunctive yet merging, circuitous routes, to repeatedly encounter its genesis as a "moso twal nan bwa" (mentioned earlier) in the unfathomable liminality of a form that unfurls in one world only to invoke another. Whether across battlefields, religions, continents, or material and immaterial realms of being, the ritual flag signals the interpenetration of domains.

However, the perpetually elusive, emergent character of the sacred *drapo* became more directly evident to me in the initial public ceremony of a *kanzo* (initiation) cycle that I witnessed in Port-au-Prince in July, 1993. As in Clotaire's origin myth (at least initially), Manbo Ayizan, the *Lwa* who protects and purifies initiates during their rites of passage in the *kanzo* cycle, presided. As the ceremony unfolded, the flags became active and stationery markers of liminality intimately involved in the dynamic of the candidates' death and rebirth into spiritually articulated beings.

The initial salutes, runs, and mock battles between the flagbearers and the *laplas* were surprisingly flat. Perhaps because the *pòt drapo* appeared reluctant, nervous, and not quite sure of their movements, this ritual sequence seemed more pro forma than visually stunning and activating. I was also jaded by reading accounts that necessarily isolated the flag presentation for purposes of description and interpretation, possibly heightening the spectacle of their initial, choreographed routines. No one standing around me seemed particularly attentive while it was unfolding, and I had to resist conversation in order to focus on what the trio was doing after the flagbearers twirled, releasing the sequined flags for the *Lwa* Sen Jak and Danbala, which had been covering their shoulders. What I had anticipated as a major spectacle seemed thoroughly absorbed into the whole: the hundreds of people pressing into the *peristil*, many of whom were dressed in white so spotless that it seemed to shimmer; the sound of their voices in talk or in songs, responding to the insistent drums and amplified by the microphones carried by the *oungenikon* (song leader) and hanging from the roof beams; and the darkness outside the *peristil* where the nocturnal promise of the military regime's random and selective violence heightened the sense of collective refuge within the temporarily illuminated enclosure whose boundaries the *drapo* now marked and consecrated.

But perhaps I simply forgot the impact of the first appearance of the flags in the emotional richness of the ceremony that grew almost unbearably intense as the group of candidates for initiation became the focus of ritual attention. This was a service to "antre yo nan djevo a" (enter them into the initiation chamber), to publicly sacralize them for a two-week period of seclusion and to bid them farewell not only because of their impending removal from the community but because their transformation would be fundamental. During this period, they would *kouche* (lie down) inside the *djevo* until their public emergence for the *boule zen* (burning of the pots) ceremony in the *kanzo* cycle. As Karen Brown explains,

> In Haitian Creole, the verb *kouche* (to lie down, to sleep, to make love, to give birth—less commonly to die) is the general word used to describe initiation. Entering the initiation chamber is like dying. Friends and family members cry as they line up to kiss the initiates goodbye. . . . As in many other sorts of initiation around the world, to *kouche* is to be forced by ritual means to regress, to become a child again, to be fed and cared for as a child would be, only to be brought rapidly back to adulthood, a new kind of adulthood, again by ritual means. (1989:272)

The initial presentation of the *drapo* marked the conclusion of the ritual sequence known as *chire ayizan* (shredding of palm leaves) and the salutation of the Rada *Lwa* who were called in succession by the songs and rum rhythms appropriate to each (see Fleurant 1996) after the flags were temporarily withdrawn. During the *chire ayizan*, a palm branch was lifted by the presiding *manbo* from a chair covered with a white sheet and positioned on the *vèvè* (ritual emblem) for Ayizan. After being presented to the four directions, its leaves were shredded by a group of *ounsi* who danced it around the *poto mitan* and into the *djevo* where the novices would be sequestered. The fringed palm leaf or *ayizan* is an emblem of the patronage of the *Lwa* Manbo Ayizan. The faces of the initiates would be covered with palm fringe when they later emerged from seclusion in the concluding ceremony of the cycle. Following the *chire ayizan* and the salutations, mock battles, and runs of the *kò drapo*, the flags were leaned against each side of the door to the *djevo* for the Ogou or warrior spirits, partially unfurled and pointing away from each other, with the sword of the *laplas* stuck into the wall between them above the door. It was explained to me that this configuration was a salute to Sen Jak, whose *vèvè* it mirrored in action.

The flags were carried into the center of the *peristil* again as the novices were brought out from under a white sheet beneath which they had been led from one of the altar rooms and positioned in a line, each one held from behind by a *oungan* or *manbo* attending them. As the *kò drapo* performed an abbreviated version of their initial routines, I was able to see the five initiates as they emerged. It was a pivotal moment for me because I saw the reality of the spiritual death and rebirth undergone in initiation already presaged in their faces, my first real exposure to the public ceremonial expression of Vodou. Four of them, young women probably in their twenties with their hair closely cropped, wore simple shirts made from coarse light grey and brown cotton. The fifth, a man who appeared slightly older, wore a sleeveless white shirt over loose pants tied with a string. Though they had not even begun to undergo the intense (and secret) rite of transformation inside the *djevo*, they already seemed to have shed their personalities and become vulnerable to the point of annihilation or transfiguration—both possibilities were visible in their faces where the anguish of approaching extinction seemed to yield a sudden, unsettled radiance shadowed in the next instant by returning fear and the will to resist. The *oungan* and *manbo* behind them continued to guide them gently or forcibly as they alternated between periods of total submission and fierce resistance. They twirled them repeatedly to the left and then to the right, periodically pressing them down to curtsey or kiss the ground in the sequence of movements used to salute a ritual superior.

After heralding the emergence of the novices, the *pòt drapo* rolled up the flags but continued to carry them during the remainder of the ceremony, blending into each ritual event. They became constant presences rather than dramatic markers, often remaining close to the novices like the *manbo* and *oungan*, attending them and mirroring their seriousness. Also, in their furled condition, they seemed to evoke the status of the initiates as embryonic spiritual beings. The *pòt drapo* accompanied a *manbo* who emerged from the altar room carrying *kanzo* necklaces—long strands of multicolored beads—which she crossed over the backs and chests of the *oungan* and *manbo* and others present who had already passed through this level of initiation. The flags then rejoined the novices who stood in a line as one by one they were slowly sacralized. Marie Yolande, the presiding *manbo*, who normally moves at an unhurried, deliberate pace as if nothing could disturb her internal rhythm, seemed to proceed in an even slower mode as she traced an asterisk like star or *pwen* in flour on the top of each initiate's head, placed a sheet over it, and then shook the *ason* (sacred Vodou rattle) above it. The five seemed remote, almost beyond contact, and the *peristil* grew quiet as the moment to bid them farewell approached. Those around me began to cry, and I joined them, feeling an awe akin to grief at the immensity of the transformation the initiates had to undergo on their path toward spiritual maturity, a transformation that had already visibly begun.

After they were embraced and kissed by the *manbo* and others present, the five were blindfolded. Those who had been guiding them stood between them, and all of them joined hands with one *pòt drapo* at each end. The line then whipped around, circling the *poto mitan* and flying back and forth across the *peristil.* The initiates began to lose their balance and then struggle wildly as if simultaneously resisting a spirit trying to mount them and their impending captivity. An *ougan* fought to keep the young man among them from breaking loose from the line. The *pòt drapo* took up positions on each side of the door to the *djevo* holding the furled flags upright. Arms and legs flying, the initiates were pushed one by one inside.

Multiple origin narratives of the *drapo* Vodou converged simultaneously in this intensely transforming ceremony. I saw the military/political protocols referencing French colonial history and the battles and heroes of Haitian Independence celebrated in the ritual deployment of the flag for Sen Jak, the Catholic processional modes suggested by the initial salutes and choreographed routines, as well as by the flags at either end of the line of initiates, and, throughout the ceremony, the spiritually activating power of flags as both stationary and mobile forms, a legacy of Kongo practices. But it was the furled flags as basic liminal markers of the regressed, inchoate yet radiant state of the initiates that evoked the most primal, and finally, for me, the most moving of all the origin myths—the "moso twal nan bwa," signaling our capacity to embody spirit, to continuously transform.

Notes

The following chapter draws on research conducted for my doctoral dissertation, "For the Flower of Ginen: the Artistry of Clotaire Bazile, a Haitian Vodou Flagmaker," Harvard University, 1998. For other published essays or interviews based on my work with Clotaire Bazile, see " 'I Am Going to See Where My *Oungan* Is," in *Sacred Possessions: Vodou, Santería and Obeah in the Caribbean* (Wexler 1997); special section on Clotaire Bazile in *Callaloo* (Wexler 1997); and "Fictional *Oungan*: In the Long Shadow of the Fetish," in *Research in African Literatures* (Wexler 2001). My deep gratitude to Clotaire Bazile, Paulette Marotière, and Alix Dabady for making the insights about the ritual flag expressed here possible and to the art historian LeGrace Benson for her devoted and inspired scholarship on Catholic and other religious/historical influences on the sacred *drapo*.

1. Thanks to Lionel Hogu for his invaluable help as a research assistant with the translation/interpretation of this passage, as well as multiple other interviews and ethnographic materials that formed my dissertation.
2. Lionel Hogu, personal communication, October 18, 1996.
3. Interview with Edgar Jean-Louis, Port-au-Prince, March 2, 1993. Edgar made these remarks during a discussion about why he thinks tourists and collectors seek out seasoned ritual flags and are willing to pay more for them than the commercial version. His statements probably reflect his appreciations of the spiritual resonance of these objects rather than what drives most collectors.
4. Edgar Jean-Louis, interview, Port-au-Prince, March 2, 1993.
5. Monique and Pollone Colin, interview, Port-au-Prince, March 4, 1993.

References

Benson, LeGrace. 1996. Some Breton and Muslim Antecedants of Vodou Drapo. In *Sacred and Ceremonial Textiles: Proceedings of the Fifth Biennial Symposium of the Textile Society of America*. Chicago, IL: Art Institute of Chicago.

Bleier, Suzanne Preston. 1995. *African Vodun*. Chicago: University of Chicago Press.

Brown, Karen McCarthy. 1989. Systematic Remembering, Systematic Forgetting: Ogou in Haiti. In *Africa's Ogun*, ed. S. Barnes. Bloomington: Indiana Univeristy Press.

——. 1991. *Mama Lola: A Vodou Priestess in Brooklyn.* Brooklyn, CA: University of California Press.

Chauvet, Marie. 1960. *Fon des Négres.* Port-au-Prince, Haiti: Edition Henri Deschamps.

Cosentino, Donald J., ed. 1995. *Sacred Arts of Haitian Vodou.* Los Angeles: UCLA Fowler Museum of Cultural History.

Dayan, Joan. 1995. *Haiti, History, and the Gods.* Berkley, CA: University of California Press.

Desmangles, Leslie G. 1992. *The Faces of the Gods: Vodou and Roman Catholicism in Haiti.* Chapel Hill, NC: University of North Carolina Press.

Fleurant, Gerdes. 1996. *Dancing Spirits: Rhythms and Rituals of Haitian Vodun, the Rada rite.* Contributions to the Study of Music and Dance, no. 42. Westport, CT: Greenwood.

Holly, Arthur. 1928. *Dra-Po, Etude Esoterique de Egregore Africain, Traditionnel, Social, et National de Haiti.* Port-au-Prince: Imprimerie Nemours Telhomme.

Hurston, Zora Neale. 1995. *Tell My Horse: Vodou and Life in Haiti and Jamaica.* New York: Library of America.

Marcelin, Milo. 1950. *Mythologie vodou, Puerto ... de la música folklórica de Cuba.* La Habana: Ministerio de Educación.

Maximilien, Louis. 1945. *Le Vodou Haitien.* Port-au-Prince, 1945.

Mennesson-Rigaud, Odette. 1958. Le Role du Vaudou dans Indépendence d'Haiti. *Présence Africaine,* 18/19 (February/March).

Métraux, Alfred. 1972. *Vodoo in Haiti.* Repr., New York: Pantheon, 1989.

Michel Philippe Lerebours. 1989. *Haiti et ses Peintres de 1840 à 1980: Souffrances et Espoires d'un Peuple.* Port-au-Prince: Imprimeur.

Mintz, Sidney, and Michel-Rolph Trouillot. 1995. The Social History of Haitian Vodou. In *Sacred Arts of Haitian Vodou,* ed. Donald J. Consentino. Los Angeles: UCLA Fowler Museum of Cultural History.

Polk, Patrick. 1995. Sacred Banners and the Divine Cavalry Charge. In *Sacred Arts of Haitian Vodou,* ed. Donald J. Cosentino. Los Angeles: UCLA Fowler Museum of Cultural History.

Rigaud, Milo. 1953. *Secrets of Voodoo.* Paris: Editons Niclaus.

Sanneh, Lamin. 1983. *West African Christianity: The Religious Impact.* Maryknoll, NY: Orbis Books.

Shannon, Magdaline. 1983. Introduction to Price-Mars, J. In *So Spoke the Uncle.* Trans. Magdaline Shannon. Washington, DC: Three Continents Press.

Stebich, Ute, ed. 1978. *Haitian Art.* New York: Brooklyn Museum.

Thompson, Robert Farris. 1993. *Face of the Gods: Arts and Altars of Africa and the African Americas.* New York: Museum for African Arts.

Thompson, Robert Farris, and Joseph Cornet. 1981. *The Four Moments of the Sun.* Washington: National Gallery of Art.

———. 1983. *Flash of the Spirit.* New York: Vintage Books.

Thornton, John. 1992. *Africa and Africans in the Making of the Atlantic World, 1400–1680.* Cambridge: Cambridge University Press.

Trouillot, Michel-Rolph. 1990. *Haiti, State Against Nation: The Origins and Legacy of Duvalierism.* New York: Monthly Review Press.

Yonker, Dellores. 1991. *Sequinned Surfaces: Vodoun Flags from Haiti.* Northridge, CA: Art Galleries California State University, Northridge.

Chapter 6

"The Jew" in the Haitian Imagination: Pre-Modern Anti-Judaism in the Postmodern Caribbean

Elizabeth A. McAlister

Europe and the Jews: A Mythological Blueprint for Demonization

Each year in Haiti the Holy Week of Easter sets the stage for spiritual dramas of remembrance, performed in carnivalesque street theater throughout the country. While Catholics reenact the Passion of Jesus and enter with him into his tribulation and resurrection, some practitioners of the Afro-Haitian religion called Vodou organize enormous musical parades called Raras and take to the streets for the spiritual warfare that becomes possible when the angels and saints remove to the underworld, along with Jesus, on Good Friday. The cast of characters who have a hand in the week's events include the deities of Vodou—especially Baron Simitye, the Vodou "Lord of the Dead"—the *zonbi* (recently dead) who are his wards, and also Jesus, the two thieves crucified with him, a couple of Haitian army officers who secretly witnessed the resurrection, Pontius Pilate and the Romans, Judas, and "the Jews." The week's ritual events combine the plots and personae of the Christian narrative with the cosmology of various African religions and rehash them in local ritual dramas whose elements draw from the entire history of the Atlantic world, from the European Middle Ages to the contemporary condition of global capitalism in the Americas.[1]

The Haitian Lenten Rara season remembers a certain history of the Americas. Said by Haitians from the town of Leogane to be "an Indian festival," the Raras provide a fleeting yearly remembrance of the 250,000 Tainos who died in the first two years after Christobal Colon's (Christopher Columbus) fateful 1492 arrival in Haiti, known as Aiyti-Kiskeya, the "mountainous land."[2] But this is only the first of many fragmented historical memories. Harnessing the spiritual power of the deities in the Petwo, Lemba, and Kongo branches of Vodou, the Raras also recall and activate religious principles from the African kingdom of Kongo that flourished in the fourteenth and fifteenth centuries. The festival carries Creole memories also, layers of American-side history. Rara parades come to their climactic finish in Easter week precisely because Holy Week was mandated (in 1685, under the Code Noir) to provide a respite from labor for enslaved Africans of the colony. Slavery and the distinctions between freedom and servitude are themes in the Raras and in Vodou, as they are in the Christian story.[3]

Besides remembrances of the Indians, the Africans, and the enslaved Creoles, however, Haitian Holy Week is also an heir to the anti-Judaism of medieval European popular thought. After all, Christobal Colon and the early colonists were products of the religious worldview of the late Middle Ages, when the Inquisition was in full force. In a telling coincidence of history, Colon set sail for what he would call the *outro mondo* (other world) in August of 1492, only three days after the final departure of the Jews from Spain.[4] This was the era during which Spain expelled its entire Jewish population, and the Inquisition reserved special tribunals for any *anusim*, or *conversos*, converted Jews, who were suspected of "Judaiizing."[5] As figures to be manipulated, demonized, or embraced, "the Jews" are marked as the original "Other" of Europe, the very first object of Christendom's projection, marginalization and demonization. Europe's demonization of Jews became a mythological blueprint for the encounter with Native peoples and Africans in the Americas.

Easter week in Haiti tells many histories, then, but it is this last one that interests us here: the demonization by European Christianity of two groups—Jews and black Africans. This work is about how some black Africans (Haitians) have inherited, used, and manipulated European Christian anti-Judaism. I will suggest here that many of the negative images of Vodou draw from and elaborate upon medieval European images of Jews. These flexible popular tropes hinge on the figure of the devil and link the devil first with "the Jews" and then, in colonial Saint-Domingue, with the Africans and Afro-Creole Vodouists.[6]

An anthropologist who specializes in Haiti, I was studying the Rara festivals throughout Haiti and came upon a series of obscure rituals and conceptions revolving around notions of "the Jews." In contemporary Haiti, local dramas replay the Christian ritual cycle of death and resurrection at Holy Week and represent the symbolic presence of Judas and of "the Jews." The way their present-day descendants portray and perform "the Jew," however, tells a complicated and ambiguous story.

The story is not simply about a one-directional process of "Othering" and demonizing a conquered people. I am also interested in the agency of the disenfranchised, in their expressions, reactions, and representations. Some present-day practitioners of Vodou manipulate the inherited, demonized images of "the Jew" in both alarming and creative ways. In the course of Easter week, "Jews" are demonized and burned in effigy by some—but they are also honored and claimed by others as forefathers and founders of the Rara bands. Various Rara leaders embrace the identity of "the Jew" and claim a sort of mystical Jewish ancestry. In accepting the label of "Jew," these Rara leaders might be understood as taking on a mantle of denigration as a kind of psychic resistance. In carving out a symbolic territory as "Jews," they symbolically oppose the powers that historically have sought to exploit them—the Haitian Catholic elite.

Through the image of Jews in the Haitian imagination, we can interpret the process of domination that married Christianization and anti-Semitism to a process of racialized capitalist expansion in the Americas. The imaginary reserved for European "demonic" Jews is portable and easily transferred into the Indian and African peoples of the Atlantic world. But myths, by their nature, create imminent and shifting imaginaries, less easily controlled by orthodoxy. Exploited peoples embraced the image of the Jew and creatively performed oppositional dramas in which they critique the morality of Christianity and their own place in the class structure.

Postmodern Peasants: Rara and Class in Haiti

The entire Lenten season is politically charged for Haitian society as the annual period in which peasant-class people are sanctioned by tradition to parade and sing. Rara begins right after Carnival on Ash Wednesday and builds throughout Lent until Easter weekend. Occurring in multiple localities, Rara's Easter week parades represent the largest popular gatherings of Haitian *pèp-la*, (the people, the folk). Groups of fifteen to several hundred people play drums and bamboo horns, dance along the roads, and stop traffic for miles in order to perform rituals for Vodou deities at crossroads, bridges, and cemeteries. Rara underscores the opposition of Vodou and Catholicism because of Rara's boisterous public presence during the period of Lent, a Catholic time of solemnity and self-deprivation.

Given the drastic disparity of wealth in Haiti, the appearance of thousands of peasant-class people in public space is inherently a deeply charged moment, considered "dangerous" both culturally and politically by dominant groups. For members of the educated enfranchised classes, hundreds of noisy people celebrating in the streets conjure an image of their nightmarish fantasies about mass popular uprising. As a large-scale popular festival, Rara is structurally oppositional to the dominant classes who make up the Haitian enfranchised minority: the literate, moneyed classes in their various aspects, who have historically depended on the Haitian army and U.S. support to maintain power.

It has long been routine to speak of Haiti as being a "divided society" comprised of two major classes: the rural "peasants" and the French-identified "elite." This cliché oversimplifies a complex historical process and the resulting heterogeneities of the various class actors in that country. The opposition also obscures the intimacy that characterizes the contact between the powerful and those they dominate.[7] It *does*, however, refer to a definite and blatant divide in Haitian society: between a politically and economically enfranchised minority and a disenfranchised, exploited majority. The root of Haitian inequality began in colonial plantation slavery and subsequent devastating economic policies after independence in 1804. Agricultural goods produced with the simplest technology by a growing peasantry were and still are taxed at customs houses and provide the bulk of government revenues. This basic scenario of an overtaxed, unrepresented nonliterate peasantry exploited by an urban bourgeoisie remains unchanged to the current time. An estimated .8 percent of the Haitian population currently owns 44.8 percent of the wealth.[8]

The classes who have historically performed Rara and who still make up the majority of Rara participants are the mass of the black peasantry and, recently, the urban poor. Not unlike other peoples of the so-called developing world, their essentially premodern peasant condition now lies embedded in the postmodern context of global capital. The Haitian nation-state is a virtually powerless entity on the international stage, and the peasantry and urban poor are caught in a system that constitutes them as the lowest link in a globalized capitalism. Perhaps the most crucial factor in upward mobility today is access to family and resources from *lòt bò dlo* (the other side of the water)—New York, Miami, or other points in the Haitian diaspora. Haitian transmigrants send home an estimated $100 million a year to families and small businesses.[9] Following the realities of transnationalism, Rara parades are also dotted with *djaspora* (literally, "diasporas") the returning townspeople from abroad who come to participate in the festival during precious vacation weeks off from work.

"Heat up the Rara": Mystical Contracts and Community

Most Rara bands are affiliated with *ounfo* (temples) of Afro-Haitian religion, as well as with secret societies called Bizango and Shanpwel. Rara can be read as an annual ritual period when the spiritual work of the *ounfo* is taken into public space. In this sense, Rara is a peripheral branch of Afro-Haitian religion known nowadays as Vodou. This religious culture of the Haitian majority consists of a fluid, inherited, oral tradition of relationships with deities from various African societies, as well as relationships with ancestors.

The major historical contributors to the creolized system called Vodou were the cultures of Dahomey, Yoruba, and Kongo, with pockets and influences from numerous other ethnic groups among the enslaved Africans who began to arrive in Hispaniola as early as 1512. When the French gained control of the territory in 1697, the French style of Catholicism of the later colony was superimposed onto the Spanish Catholicism of the early period. The Africans re-created their beliefs and practices in a complex process of creolization, embracing parts of Catholicism as well as elements of Freemasonry, French occultism, and African Islam.[10]

Vodou is concerned most fundamentally with the healing arts, in physical, psychological, and spiritual realms. Anthropologist Karen McCarthy Brown convincingly argues for the centrality of healing to Vodou and that healing must in turn be understood in terms of a cultural definition of personhood. As individuals are defined by their relationships both to ancestors and to those living in the community, illness is always approached through a careful consideration of imbalance in relationships. Treatment is effected by attending to the webs of relationships that define a person in the Haitian context. The spiritual work of Vodou has to do with *chofe* ("heating") life energy and restoring it to balance.[11] There is also an established set of practices related to the forcible manipulation of relationships with others. Vodouists and anthropologists alike distinguish between a *fran Ginen* (literally, "true African") moral system and the immoral practices of the *boko* (sorcerer).[12]

Rara bands parade through public thoroughfares using music, song, small bonfires, and other ritual techniques of Vodou, in order to *chofe* relationships with Vodou *lwa* (deities) who protect and direct the bands. Most Rara bands are under the patronage of a Vodou *lwa*, and participation in a band is a form of spiritual service. Bands visit local cemeteries and serenade the ancestors, sometimes enticing any interested *zonbi* (souls) of the recently dead to join the Rara in a spiritual attachment that will *chofe* the drummers and mystically work to *kraze* (crash) other neighborhood bands. Rara bands think of themselves as small armies, out on maneuvers to perform mystical work and carve out territory. The bands are often *angaje*, or under a mystical contract to parade for seven years in a row. This serious spiritual engagement strengthens the relationship of the whole community with the *lwa* of Vodou and enhances good fortune and health.[13]

Bwile Jwif: "Burning the Jew" in Effigy

It was Holy Thursday night in 1993, and my research team and I were out recording and filming a Rara band in the narrow back-streets of Port-au-Prince. We were dancing along down the dark hilly streets at a good clip, on our way to a small cemetery to try to get some *zonbi* to *chofe* (heat up) the band for the season's climax on Easter. We stopped while the band paid a musical salute to the invisible guardian of the cemetery gates in Vodou. I looked up and noticed a straw dummy sitting on the roof of the house across the street. It was a "Jew."

He was sitting in a chair in the open air, on top of this one-story tin-roofed house. Made of straw and dressed in blue jeans, a shirt, suit jacket, and sneakers, this "Jew" wore a tie and had a pen sticking out of his shirt pocket. His legs were crossed, and over them sat what looked to be a laptop computer fashioned out of cardboard. A cord seemed to run from the computer down into a briefcase that sat by his chair.

I asked around for the *mèt Jwif-la*, its owner. An older man missing a few teeth came forward, offering a calloused, muscular handshake that revealed a life of hard physical labor. He was from the countryside in the South of the island, a migrant to Port-au-Prince. I found myself in the ridiculous position of having to complement him on his work. "Nice Jew you've got there," I said, "*Ou gen yon bèl Jwif la, wi.*" "Oh yes, we leave it up for the Rara band to pass by. Tomorrow afternoon we'll burn it," he said. "Aha . . . well . . . great . . ." said my research partners and I, flaring our eyes at each other. I guess nobody told the guy that Jean-Claude Duvalier banned the practice in the 1970s, around the time of a rush of tourism and foreign industrial investment. I bet other people still do it, here and there.

The Easter ritual of burning "the Jew" or burning "Judas" in effigy was practiced until recently by all classes in Haiti. There were many local variations, but usually by Maundy Thursday an effigy was erected in some central location, and at 3:00 on Good Friday it was burned by the local community.[14] This was done in a ritual retaliation against Judas, who betrayed Jesus, or against "the Jews" who "killed Jesus."

The symbol of the Jew surfaced around Easter among all classes in most regions of Haiti. Local peasant communities enacted this carnivalesque theater and so did wealthy plantation households. Thérèse Roumer, a writer from the provincial city of Jérémie, remembered the *Juifs errants*, the "wandering Jews" of her childhood. Her father owned expansive tracts of land in the region and maintained a large family home. A "Jew" was erected at the beginning of Lent. He had stuffed pants and shirt, with a pillow for a head, and he sat in a chair on the veranda by the front door. The idea, said Madam Roumer, was to kick the Jew whenever you went in or out of the door, "say any bad words you had," and scold him for killing Jesus. On the Saturday morning before Easter all of the children from town would find wooden sticks, come to the house to beat him, and then burn him up in a bonfire.[15] Children were exhorted by the grownups to "pray for the conversion of the Jews."[16] The family would then go off to church for some holy water and wash down the verandah.[17]

Most of those interviewed remembered that the Jew in effigy was part of a child's game, in which the "Jew" represented Judas himself and was hidden by the adults in the neighborhood. William Seabrook, whose book *The Magic Island* has sustained many critical blows since its publication in 1929, wrote this tongue-in-cheek account, worth reproducing in its entirety:

> On the last bright Easter morning which I spent in Port-au-Prince—this was only a year ago—the Champs de Mars, a fashionable park adjacent to the presidential palace and new government buildings, resembled an untidied battlefield on which scenes of wholesale carnage had been recently enacted.
>
> It was impossible to drive through it without swerving to avoid mangled torsos; it was impossible to stroll through it without stepping aside to avoid arms, legs, heads, and other detached fragments of human anatomies.
>
> It was impossible also to refrain from smiling, for these mangled remains were not gory; they exuded nothing more dreadful than sawdust, straw and cotton batting. They were, in fact, life-sized effigies of Judas and Pontius Pilate's soldiers—done to death annually by naive mobs bent on avenging at this somewhat late day an event which occurred in Palestine during the reign of Tiberius.
>
> . . . I had made the acquaintance, so to speak, of one Judas before he betrayed our Lord and fled to the woods. All the little community had contributed toward his construction. He sat propped in a chair outside the doorway. They had stuffed an old coat, a shirt, and a long pair of trousers with straw, fastened old shoes and cotton gloves, also stuffed, to the legs and arms, and had made ingeniously a head of cloth, stuffed with rags, with the face painted on it and a pipe stuck in its mouth. They introduced me to this creature very politely. They were rather proud of him. He was Monsieur Judas, and I was expected to shake hands with him. You see—or perhaps you will not see unless you can recall the transcendental logic which controlled the make-believe games you used to play in childhood—that Judas had *not yet* betrayed Jesus. He was, therefore, an honored guest in their house, as Peter or Paul might have been.
>
> And so their righteous wrath will be all the more justified when they learn on Saturday morning that Judas has turned traitor. Then it is that all the neighbors, armed and shouting, the men with machetes and *cocomacaque* bludgeons, the women with knives, even more bloodthirsty in their vociferations, invade the habitation where Judas has been a guest, demanding, "*Qui bo' li?*" (Where is the traitor hiding?)

> Under the bed they peer, if there is a bed; behind doors, in closets—I happened to witness this ceremony in a city suburb, where they do have beds and closets—while members of the household aid in the search and make excited suggestions. But nowhere can Judas be found. It seems that he has fled. (What has really occurred is that the head of the house has carried him off during the night and hidden him, usually in some jungle ravine or thicket close on the city's edge. Judas usually takes to the forest as any man would, fleeing for his life. But this is not always predictable. A Judas has been known to hide in a boat, in a public garage yard, even under the bandstand in that Champs de Mars whither so many of them, wherever found, are dragged for execution.)
>
> So tracking Judas becomes a really exciting game. A group collects, shouting, beating drums, marching in the streets, racing up side-alleys; meeting other groups, each intent on finding the Judas planted by its own neighborhood, but nothing loath to find some other Judas and rend him to pieces *en passant.* Crowds may be heard also crashing and beating through the jungle hillsides. It is rather like an Easter-egg hunt on a huge and somewhat mad scale.[18]

Other cultures practice the tradition of burning Judas in effigy at Easter week, notably in Mexico and other parts of Latin America.[19] The practice may stem from the liturgical dramas, or "evangelizing rituals," practiced by early Jesuit missionaries to the Americas. The Jesuits are known to have staged elaborate dramas in the communities where they worked, playing out scenes from Jesus' life.[20] Passion Plays spread the idea of Jews as "Christ-killers." According to this ritual logic, Judas, who betrayed Jesus, is conflated with "the Jews" who "mistreated Jesus," making all Jews into "Judases."[21] The supposed role that the Jews played in the Crucifixion, as described in the New Testament, embellished in legend, and portrayed on the stage, was familiar to both cleric and layman. It was a good starting point for moral teaching.

The idea that "the Jews killed Jesus" is rooted, of course, in the New Testament, which can be read as a polemic that displays the anti-Judaism of the early Church. Sander Gilman has argued that the negative image of the difference of the Jew found in the Gospels (and especially, we might note, the figure of Judas) became the central referent for all definitions of difference in the West.[22] During the medieval period European Christianity produced the image of "the demonic Jew," an inhuman creature working directly for Satan. Joshua Trachtenberg writes in his classic work *The Devil and the Jews* that "the two inexorable enemies of Jesus, then, in Christian legend, were the devil and the Jew, and it was inevitable that the legend should establish a causal relation between them."[23] By the medieval period the devil was cast as the master of the Jews, directing them in a diabolical plot to destroy Christendom.

In the medieval Passion Plays that set the tone for the popular Christianity of Christobal Colon's Europe and the colonial Jesuit missions, the Jews are handed the entire weight of blame for Jesus' death, and Pontius Pilate and the Roman participants in the narrative fade into the background.[24] Medieval European Mystery Plays were popular liturgical dramas, reenacting various scenes from scripture. They grew into village festivals performed in marketplaces and guildhalls, taking on the "secular, boisterous, disorderly and exuberant life of the folk."[25] In *Le Mystere de la Passion*, a fourteenth-century French play depicting the Crucifixion, the Jews are the villains of the piece, egged on by devils. In the climax the devils instigate Judas to betray his master, and they howl with glee when they are successful.[26] The idea of Jews as demonic

"Christ-killers" is enhanced throughout the medieval period, forming a central theme of anti-Judaism that will authorize the persecution of Jews during the Inquisition.

The clergy of the Spanish period, like the French that followed them, were small in number and faced the overwhelming project of establishing and maintaining Christianity. It is likely that the Jesuit, Dominican, and Franciscan missionaries in Hispaniola made use of the theatrical tactics deployed by their colleagues in New Spain to convert the Native Americans. In that colony, large-scale popular dramas were modeled after the Mystery Plays of Spain and France, depicting the winners and losers in the Christian story and making clear parallels to the colonists and the conquered. Judas, "the Jews," Jesus, and the apostles made for casts of characters that would illustrate the larger drama of power relations at the start of the colonial enterprise.[27] The Christian story and theatrical public rituals generated narratives meant to authorize and display the technologies—chains and whips—of servitude. European Christendom dramatically set itself up as a sole civilizing force against the barbaric and demonic forces of Jews, Native Indians, and Africans.[28]

The historical antecedents of the Haitian *bwile jwif* ("Jew burning") rituals may well be in these sorts of Passion Plays that referenced the events of the Spanish Inquisition. In the late fifteenth and sixteenth centuries—as the Spanish were establishing the slave trade to the colonies—*conversos* believed to have secretly practiced Judaism were sentenced to be burned alive in Spain. Jews in hiding were sentenced in absentia and burned in effigy.[29] These auto-da-fé practices were likely the model upon which the Latin American rituals are based. Although the Inquisition was never organized in Hispaniola, the Easter effigy burnings are most likely rooted in Inquisition symbolism and its attendant public ritual terror.[30]

Other bits of cultural flotsam and jetsam may have trickled down from Inquisition history. The *lwa* Papa Gede, in his own code language, calls the pig "Jwif." Surely Papa Gede is remembering one of the most common caricatures of the Jew in the Middle Ages, the notorious figure of the *Judensau*, in which a sow feeds her Jewish offspring with the devil looking on.[31] Perhaps the expression is an inverted survival of the fifteenth-century term for the Spanish *conversos*, who were called *marranos* (swine; pig) after the Christians conquered the Moors.[32]

Anti-Jewish sentiment was an implicit part of the ruling process of the French colony of Saint-Domingue. The Church itself was among the largest of the slave-owning landholders in the colony, and it won an advantage with the establishment of the Code Noir.[33] This edict by King Louis XIV mandated the planter class to baptize and Christianize the slaves, just as it simultaneously outlawed the exercise of any religion other than Catholicism. The Jesuits, working as an order before the 1704 official establishment of their mission, manifested a marked dislike of Jews and their religion. In 1669 they appealed to the Crown representative to take actions against "tavern keepers, undesirable women and Jews."[34] In 1683 the Church induced King Louis XIV to expel all Jews from the colony and to impose a religious test on new immigrants.[35]

It would have been only logical, then, for the colonial clergy to take the image of the Jews as an evil, anti-Christian force and hold them up in comparison with early forms of Vodou—the real threat to Christianity in the colony. Although the Christianization of the Africans in colonial Saint-Domingue was a halfhearted and badly organized enterprise, enslaved people were mandated by the Code Noir to be baptized, and they sporadically attended mass, married, and were directed in catechism.[36] In their efforts to control the enslaved, the clergy preached Paul's letters to the Ephesians and other biblical passages exhorting slaves to obey their masters.[37] Most of their practical worries revolved around the "superstition" of the Africans, their magical abilities, and their knowledge of poison, for greater than the fear of diabolism was the more imminent threat of uprising and rebellion. Numerous regulations were passed in the colonial period and after, making various religious and magical practices illegal. Underlying anti-Vodou sentiment was the notion that the Africans, like the Jews before them, were acting in consort with the devil.

The litany of charges that were leveled against Jews in medieval Europe were transferred wholesale onto the Vodouist. The list of devilish crimes that were attributed to European Jews was an elaborate series of evil activities aimed at destroying Christendom. Jews were accused of a range of magical crimes, from superstition, sorcery, and desecration of the host, all the way to ritual murder, the drinking of Christian blood, the eating of human flesh, and poisonings.[38] It is striking that this list is replicated in the colony, targeting Africans and Creoles of Saint-Domingue.

Like the *marranos*—converted Jews constantly under suspicion of "Judaizing"—African converts to Christianity were suspected of sorcery. Joan Dayan writes of the late eighteenth century, "It seemed as if the more Christian you claimed to be, the more certainly you could be accused of conniving with the devil."[39] A decree passed in 1761 complained that slaves' religious meetings at night in churches and their catechizing in houses and plantations were actually veiled opportunities for prostitution and marronage. Slaves who had taken on roles of "cantors, vergers, churchwardens, and preachers" were charged with "contamination" of sacred relics with "idolatrous" intentions.[40] Africans requested to be baptized over and over, believing in the mystical properties of the rite, or wishing to attend the accompanying feast.[41]

The legal codes from the colony to the present criminalize numerous practices of sorcery, linking the devil with the Africans and Creoles. A decree passed in 1758, for example, prohibited the use of "*garde-corps* or *makandals*."[42] Still in use today as *pwen* (literally "points"), these "body-guards" were objects infused with spiritual force, directed to protect their wearers. Makandal was also the name of the famous maroon leader in the Haitian revolution. An adept botanist as well as a revolutionary, Makandal was convicted of instigating a campaign of poisoning planters' wells in 1757, during which more than 6,000 whites were poisoned.[43] Besides being labeled superstitious, sorcerers, poisoners, and false Christians, Africans and Creoles were accused of stealing and desecrating the host, drinking blood, and cannibalism, charges that rounded out and replicated the litany of anti-Christian charges against Jews.

Satan's Slaves: Vodouists in the Catholic Imaginary

Throughout Haitian history, the Catholic clergy and the enfranchised classes have cast Vodou as a cult of Satan, a complex of African superstitions to be purged from the beliefs of the Haitian majority. In cycles of violent repression, Vodou practitioners have been jailed, tortured, and killed, and sacred objects have been burned. Using the image of slavery so salient to a population once enslaved and perpetually negotiating its sovereignty, the Church's antisuperstition campaigns targeted Vodouists as slaves of Satan, who is himself working to contaminate and destroy Christianity. Consider this rhetoric from a Haitian catechism of the antisuperstition campaigns of the 1940s:

> —Who is the principal slave of Satan?
> The principal slave of Satan is the oungan. [Vodou priest]
>
> —What names do the oungan give to Satan?
> The names the oungans give to Satan are the lwa, the angels, the saints, the dead, the twins.
>
> —Why do the oungan take the names of the angels, the saints, and the dead, for Satan?
> The oungan give the names angels, the saints, and the dead, to Satan to, deceive us more easily.
>
> —Do we have the right to mix with the slaves of Satan?
> No, because they are evil-doers and liars like Satan.[44]

In Haitian cultural politics, Catholicism has positioned itself against Vodou as an official, European, legitimate, orthodox tradition associated with ruling power and authority. Vodou occupies an oppositional space which is Creole, home-grown, unorthodox, diverse, and by extension illegitimate, impure, evil, and satanic. Politically, then, the two traditions have been constructed as polar opposites. The Lenten period becomes an interesting and tense time during which Catholic and Vodou practices clash. The performance of Rara during Lent, within the Roman Catholic yearly calendar, reveals its historical evolution as a festival celebrated in a world dominated by Catholicism. The Rara festival unfolded in an Afro-Creole cultural space juxtaposed against a Catholic order, and its performance each year underscores the political oppositions between the two symbolic systems.

The political uses each tradition makes of the other is only the most public face of culture, however, and obscures the complex interactions between the traditions, the ways they combine themselves, and the ways individual people combine them in practice. Focusing on their political opposition obscures the dialectic figuring and reconfiguring inherent in historical processes of creolization. Writing on Afro-Cuban religion, David H. Brown points out that "An overemphasis on 'religion,' the binary positioning of 'African' and 'European/Catholic' systems, and the stark racial opposition of 'white' and 'black' limits our comprehension of the multiplicity of experiences, influences, and roles Afro-Americans *chose* in complex Caribbean creole societies. 'African' and 'European' interacted less as static capsules than as historical processes."[45]

American cultures evolved through processes of creolization, wherein cultural tropes and symbols shift and reconfigure themselves within unequal power relations. Both the Afro-Haitian religion and the Catholicism that evolved in Haiti were constructed in dialectical relation to each other. To a significant degree, Vodou and Catholicism each has incorporated the other into its philosophies and practices. Each tradition is constitutive and revealing of the other.

Cultural complexes that evolve in unequal relations of power take on a process similar to the culture wars between "high" and "low" culture articulated by Stallybrass and White:

> A recurrent pattern emerges: the "top" attempts to reject and eliminate the "bottom" for reasons of prestige and status, only to discover, not only that it is in some way frequently dependent upon the low-Other . . . but also that the top includes that low symbolically, as a primary eroticized constituent of its own fantasy life. The result is a mobile, conflictual fusion of power, fear, and desire in the construction of subjectivity; a psychological dependence upon precisely those others which are being rigorously opposed and excluded at the social level.[46]

Institutional Catholicism depends on its opposition to Vodou, for it is its position against what is impure and illegitimate that strengthens Catholic virtue in Haiti. In the Christian story, the trope of the Jew is used by the enfranchised classes as a kind of fantasy "low-Other" that authorizes Catholic bourgeois superiority. The equation of non-Christians with Jews gave bourgeois Haitians one more cultural difference between themselves and the nonliterate Vodouists. Besides being dark-skinned, nonliterate, Creole-speaking peasants, they also were pagans, and anti-Christians. Symbolically, they were Jews. Haitian Catholics came to depend, in a sense, on the trope of the Vodouist-Jew as a force to oppose and exclude, a way to define the Catholic self through a negative referent.

The myths and rituals that surface at Easter yield particularly illustrative readings for the way in which both "high" and "low" culture groups reach for symbols and embrace, perform, and transform them in the ongoing process of negotiating power. The performances of Easter myths range from the strictest Catholic mass to the popular Easter rituals sanctioned by Catholicism, all the way to the oppositional readings of Rara bands.

Theologically, Easter is the most important holiday in the Catholic calendar, celebrated in Haiti both in official church mass and in popular rituals. One of the Easter traditions practiced by all classes is the reenactment, after church on Good Friday, of *Les Chemins de la Croix*, the stations of the cross. For this Passion Play a series of ritual stations are set in place, and barefoot pilgrims, some dressed in burlap, visit each station, fasting, without water, reciting prayers before each spot. A local man plays the role of Jesus, and other actors portray other figures in the story. The Passion Play was honed as a genre in medieval Europe, and this somber drama drawn from the four gospels still enacts itself in numerous locations on Good Friday all over the Christian world.[47]

At the same time that Catholics engage in these Easter rituals, Rara bands are busy parading through public thoroughfares. In fact, some Raras deliberately plan to walk past churches on Sunday to annoy the Christians. In 1993 a priest in Pont Sonde ended mass with the admonishment, "Don't go in the Rara," worried he might lose some parishioners to this "devil's dance." In the imaginary of the Haitian bourgeoisie, Vodouists have been cast as evil slaves in Satan's army. As anti-Christians, they became symbolic Jews.

"If you go in the Rara, you are a Jew."

A Rara band called "Ya Sezi" ("They will be surprised") walked for miles all day on the Good Friday of 1993, along the banks of the Artibonite River. They were on their

> way to the compound of Papa Dieupe, a wealthy landowner in the region, and also the "Emperor" of a Shanpwèl society. My team and I had chosen Papa Dieupè's as the best place to be for Rara; we figured we could comfortably stay put in one place and watch the bands come to salute the "big man."
>
> Ya Sezi's entrance was spectacular for a sleepy country day. We could hear the *banbou* blowing for miles, and children would run through and breathlessly announce that the band was coming to salute the Emperor. They came up the path, and did the ritual salutes for the Vodou spirits living in the trees in the compound, and then turned to salute Papa Dieupe's "children" in the society. Finally, after they'd played until about midnight, Papa Dieupe himself emerged from his small house and received them.
>
> After playing music in the compound for much of the night, the group slept, and awoke early Saturday morning to play and "warm up" before they left. While the musicians played, each of the dancers (who were all women) took turns holding the whip belonging to the leader, and ran in circles through the compound. The other dancers set off in hot pursuit, their dresses streaming out behind. Papa Dieupe told me they were taking turns being Jesus, running from the "Jewish soldiers." Pilate's Roman soldiers were nowhere in evidence, but rather had been collapsed into a new bloodthirsty figure of "Jewish soldiers." Comically enacting Jesus' suffering on his walk to Calvary, the Rara members were amusing themselves by portraying both Jesus and his "killers," "the Jews."[48]

During the Easter Rara festivals, the story of Jesus' life and death replays itself in the churches and streets of the country, and Jesus, Judas, and the Jews join the spirits of Vodou as dramatic characters to be performed and interpreted. Good Friday in particular becomes a day of stark contrasts between the *fran Katolik*, who pray, fast, and walk the stations of the cross, and the Rara bands, who parade noisily through the streets singing and working to *chofe* ("heat up") relationships with Vodou spirits and the recently dead. Catholic Haitians make a clear connection between the exuberant celebrations of Rara on the anniversary of Jesus' death, and "the Jews who killed him." A popular expressions is, *Ou al nan Rara, se Jwif ou ye*, "If you go in the Rara, you are a Jew."

Haitian Catholicism equates Vodou, the devil, and "the Jews who killed Christ," and we can see how celebrating Rara in the streets on the day Christ died "makes you a Jew" in the Catholic view. Even some university-educated Haitians have a vague concept that "Rara is a Jewish festival." At a fancy cocktail party in the wealthy enclave above Petionville I was introduced to a young Haitian architect from the mulatto class. "Studying Rara?" he asked incredulously. "Well, you'll find that it's a Jewish thing." Pressed on how a Jewish festival could have found its way through history to be adopted by the Haitian peasantry, the man shrugged his shoulders and reached for his rum punch.

Every Rara member I interviewed, on the other hand, remembered that Rara "came from Africa," with the slaves. This seems a clear historical fact: Rara continues and extends a number of African cultural principles, including the centrality of community enterprise, relationships with the ancestors and the deities, a kind of politics of "big-man-ism," the use of natural sites for spiritual work, as well as the performative African-based drumming, call-and-response singing, and dance in public festival.

After establishing the African roots of the festival, however, Rara leaders would invariably go on to articulate the idea that Rara was linked to the Jews. Many of them cited the precise origin of Rara as the celebration of the Crucifixion itself. "It was the Jews who crucified Christ who made the first Rara." One *oungan* explained it this way:

> Long ago, after they finished nailing Jesus to the cross, the soldiers who did that saw that it would be even more satisfying to put out a Rara to show that they were the winners. They put out a Rara, they made music. They were rejoicing, singing and dancing.[49]

This idea that "the Jews who crucified Christ" rejoiced and made the first Rara was related to me over and over by Rara members. The historical genealogy of the notion is obscured here, as is the cultural history of most dispossessed groups. Yet one returns to the Passion Plays of the early church, modeled after the ones in medieval France, England, and Germany. The Jews are the central villains of these stories and are directed by demons and devils hovering in the background. Together the devils and the Jews convince Judas to betray his master, and celebrate when they are successful. Joshua Trachtenberg describes it thus: "Around the cross on which Jesus hangs the Jews whirl in a dance of abandon and joy, mocking their victim and exulting in their achievement." This explicit scenario of a crucified Christ surrounded by joyful, dancing Jews celebrating their victory makes its way from the popular European imaginary to become a memory of former African slaves.[50] Another Rara president reiterates:

> Rara is what they did when they crucified Jesus, on Good Friday. At that point, all the Jews were happy. They put the Rara out, they masked, they danced, they dressed in sequins, they drank their liquor and had fun.[51]

The link between Rara and "the Jews who killed Christ" was strong enough in the Haitian imagination that Rara members became Jews in their own rememberings. A *oungan* told me that "It was the Jews who came with this tradition. Now it's become our tradition."[52] Another *oungan* provided an explanation that implicitly described how the Africans could have inherited this celebration of the ancient Hebrews. "Rara is something that comes from the Jewish nation. So, mystically speaking, Haitians are descended from Africa. The Africans always kept their mystical rites."[53] In this logic, Africans are equated with the Jews of antiquity, and it is this linkage that explains how Haitians have inherited Rara from the Jews. Through Rara, these Haitians embraced the subversive identity of "the Jew" and see Jews somehow as forerunners of their African ancestors.

When Rara members embrace the negative cultural category of the Jew, the mythology they generate may be understood as a repressed people's subversion of the ruling order. This class resistance to Catholic hegemony is a form of theatrical positioning on the part of the peasants that says "We are the Jews, the enemy of the French Catholic landowners." Like other groups that take on the negative terms ascribed to them by the powerful, Haitians take on a mantle of denigration in the face of a hostile dominant class. Just as "high culture" includes "low culture" symbolically in its self-construction, so here does the "popular culture" include the "elite" in its turn.

Laënnec Hurbon understands this dynamic historically as a creative appropriation of cultural goods:

> [the slaves'] diverting of Christianity to their own ends . . . had nothing to do with the systematic denial of Christianity, nor was it a sign of inadequacy of evangelization, but a process of making off with those elements of Christianity which could be useful in the struggle and in the construction of their new culture.[54]

Vodouists' interpretations of biblical stories can be understood as creative subversions of official discourse. Like the Rastafari of Jamaica, Vodouists are adept orators and creative interpreters of myth and Scripture. Every imaginative Vodou practitioner may offer a new visionary interpretation of Bible and of history. These versions allow Vodouists and Rara members to authorize their own history while positioning themselves, for themselves, in terms of the dominant class and its religious ideology.

Jesus Christ is the subject of much theorizing on the part of Vodouists. In one myth, God created the twelve apostles just after he created the earth and the animals. The apostles were rebellious and challenged God. In punishment, God sent them to *Ginen*, the mythical Africa of Vodou's past and future. The apostles and their descendants became the *lwa*, while a renegade apostle who refused to go to *Ginen* became a sorcerer and took the name Lucifer.[55] Throughout the oral mythologies of Vodou is a clear theme of morality and a distinction between working with the Ginen spirits and working with the forces of sorcery. Usually the sorcerer is also a slave master of captured spirits and souls, and so themes of morality are bound together with philosophical issues of slavery and freedom.

One story I was told creatively posits Jesus as the first *zonbi*, a soul that has been captured and sold in order to work for its owner. Although I have written of this elsewhere, it bears reiterating here. This myth positions Jesus and God as the innocent victims of two unscrupulous Haitian soldiers who secretly witnessed the resurrection. It was related to me by a sorcerer who confided that he knew the techniques of capturing the spirit of the recently dead (*zonbi*) and ordering it to work:

> The whole reason that we are able to raise people after they die goes back to when they crucified Jesus Christ. Christ was sent by Gran Jehovah, by *Gran Mèt* [literally "Great Master"]. He also sent Mary Magdelene, along with two bodyguards for Jesus from the Haitian Armed Forces. When Jehovah gave the password to raise up Jesus from the dead, the soldiers stole the password and sold it. It's been handed down from father to son, which is how I could get it.[56]

Vodou takes what it can use theologically and constantly re-creates itself with fresh material. The Vodouist fits biblical figures into an already existing Afro-Creole scheme. Jesus is problematic for the Vodouist: the heavy catholicizing of the French and, later, the Haitian elite makes Jesus the god of the dominant classes. This story subtly acknowledges the teller's opposition to Christianity: a worker (a Haitian footsoldier) stole something from Jesus (the god of the elite). The stolen knowledge now becomes a tool for the subordinates, since it is Vodouists who now control the resurrection secrets of God. This tale illustrates how the Vodouist uses oppositional mythology as one of the ongoing weapons in everyday Haitian class warfare.

Rara leaders I interviewed accepted the Catholic label of pagan, Satanist, and Jew and theorized their position in a specific Vodou theology. This view agrees that Rara is anti-Christian. As one leader explained, "Rara is basically against the power of God. Because Rara is what they did when they crucified Jesus, on Good Friday." This view understands Rara to be "against the power of God," in Catholic terms.[57]

On some level, however, God has abandoned Haitians. The president of *Rara Mande Gran Moun* in Léogane explained: "God made the king Lucifer. God commands the sky, and the king Lucifer commands the earth. Everybody who is poor on this earth is in hell."[58] In this interpretation, God rules the heavens but has given Lucifer control over the earth, and humans are actually the political subjects of the king Lucifer. In the face of a class structure divided by access to the means of production but marked in many ways by religious affiliation, the response of the Vodouist is to embrace and creatively rework the identity given them by Catholics.

Commenting directly on the suffering generated by extreme economic exploitation, the figure of Lucifer stands as a kind of moral commentary on the state of Haitian government and its history of class inequality.

Rara leaders construct theology through the appropriation of "high" cultural elements into allegories of empowerment. The stories of the "Jewish Rara" and the "*zonbi* Christ" construct a sort of engagement with the texts of the Catholic dominant classes in which the power of the Vodouists or Rara members is hidden inside the images of demonization. Haitian sorcerers construct themselves as active enemies of the Catholic order, as Jews, or as allies of thieves who stole from God. The narratives support Hurbon's statement that "In the eyes of the Voodooist, his mysticism is his power. Thus it may be correct to say that the Voodoo cult, since its inception with a Creole coloration, is used by Voodoo believers as a power base from which to deal with the power elite."[59]

These myths can be seen as antihegemonic counternarratives that reconfigure histories and genealogies to cast power with the popular classes. It is a common result of repressive contexts that cultural expression will generate double-voiced, allegorical strategies so that the dominant culture is turned back on itself, transformed by the subordinate. The myths generated and performed by Rara reveal how "high" Catholic culture and the "low" Vodou culture are constructed in relation to one another, each mystically exoticizing the other in the ongoing performance of class in Haitian society. Each end of the class spectrum reaches for the figure of "the Jew" to authorize its own power in the imaginary of Haitian class warfare. "The Jew" in Haiti remains largely a figure constructed from the leftovers of medieval Christianity and sustained through Catholic popular culture. Inherited by Afro-Haitians, "the Jew" is creatively re-presented as a figure allied in opposition to the Church, to the landowners, and to the Franco-Haitian elite.

Conclusion

A popular expression of surprise in Haiti roughly corresponds to the phrase "I'll be damned!" It says, simply, *Jesu, roi des Juifs!* "Jesus, king of the Jews." "The Jews" are a stock figure in Haitian popular culture, inherited in the process of Catholic European missionizing that was part and parcel of the Latin American plantation enterprise. A figure used at once as scapegoat and mystical forebear, "the Jew" can also be a comedian who speaks the unspeakable. He shows up in Carnival as *Papa Jwif,* a *Juif errant* (wandering Jew) who delivers satirical political commentary or enacts problematic issues in the community. In Port-au-Prince during the coup that ousted President Aristide, Papa Jwif was both a signal of the AIDS pandemic and a symbol of the corrupt military rulers, diseased beyond redemption. He showed up as a carnival character dying of AIDS. He was surrounded by an entourage of doctors perpetually treating him with useless remedies, coded as U.S. political forces propping up a violent and corrupt regime.

"The Jew" and Judas are most often negative markers, and to be a *jouda* (Judas) is to betray one's friends through gossip. I have heard particularly violent army officers or tonton makoutes described in low tones as *yon Jwif* (a Jew) in their cruelty or barbarism. To be greedy or stingy is to be *kras pase Jwif* (cheaper than a Jew). While most of the negative images of the Jew center on the premodern anti-Judaism focusing on Jews as betrayers and "Christ-killers," the anti-Semitic vocabulary of Jews as hoarders and usurers has crept secondarily into the Haitian cultural vocabulary.

In Vodou, "the Jew" represents a particularly potent magic, centered on the figure of Moses. Haitians have canonized Moses as a Vodou spirit of their own, and handmade ceramic figures of *Sen Moyiz* (Saint Moses) clutching the tablets containing the Ten Commandments sit on the occasional altar. With Moses long pictured in popular Christianity as the most famous magician of all time, who transformed serpents into staffs and parted the Red Sea, his magic intrigues Haitian mystics. His magic and the magic of "the Jews" in general is an attractive source of power for disenfranchised Vodouists.

All of the myths, symbols, and rituals centered on the Jew raises the question: What was the historical Jewish presence in Haiti? While Jews never established a lasting community, it is nevertheless possible to discern a thin strand of Haitian Jewish history. It starts with the genesis of the modern Americas: at least one known recently converted *converso* was aboard Cristobal Colon's ship in 1492, and five others are suspected by historians.[60] Although the forces of the Inquisition excluded Jews, Moors, and other non-Christians from the colonies, "Jews slipped through and managed to live unmolested in loosely organized communities."[61] Most colonial Jews were Sephardim, Iberian Jews of Spanish or Portuguese origin. They often came under false identities, many of them to Hispaniola, which was settled first.[62] Some came directly from France, but others made a circuitous route from initial settlements in Dutch territories, or from Spanish and Portuguese colonies after they established the Inquisition.[63] In a study on Jews in Saint-Domingue, Zvi Loker has located Jewish settlements in four zones of Haiti, including 80 Jewish families from Curaçao who settled as traders in Cap François and brought with them a prayer leader.[64]

The relationships between most Christians and the Jews in Saint-Domingue were friendly, and Jews became a subgroup of the planter and business class. As a result, the efforts of the Church to create an anti-Jewish popular feeling were not too successful. The demonized images of the biblical Jews do not seem to have been converted into explicit anti-Jewish violence.

In the late nineteenth century, Ashkenazi Jewish families arrived in the country. The pattern for these arrivals and for the descendants of colonial Jews was to assimilate and convert to Catholicism, although many today acknowledge their Jewish ancestry. There were also Jews among the Middle Eastern diaspora of the early twentieth century, who settled in Haiti to become known as *Siryen*, "Syrian," regardless of their nationalities as Syrian, Lebanese, or Palestinian.[65] Later, during the Holocaust, some French and German Jews made their way to Haiti on steamship. While most moved on to North America or Israel, a handful stayed in Haiti to live out the rest of their lives. In perhaps the most delightful symbolic reversal of all, the Haitian legislature in the 1930s declared all Jews to be of African ancestry, since they came from Egypt at the time of the Exodus with Moses. This justified permitting European Jews to settle in Haiti as enfranchised citizens, since the Haitian Constitution had made land ownership possible only to those of African descent.

Despite the small but constant Jewish presence in Haiti, there is no evidence that a synagogue ever existed. Only one Jewish cemetery was established, long ago in the colonial period; it has long since been abandoned.

As Haitians spread further abroad in their own diaspora, the founding of the State of Israel and subsequent gathering of Jews have given Jews a positive image. The Aristide government contracted for a study of Israel and its politics of returning citizenry, viewing Israel as a possible model for recouping the human potential lost in the brain drain of outmigration.

Despite the recent positive valence given to Jews in Haitian thought, the original anti-Judaic tropes of Christianity remain. As this essay has rehearsed, Easter season has always set the stage for the resurfacing of the image of "the Jew," both in popular theater and in the official church. Throughout Christendom, references to Jews were most numerous in sermons delivered during Easter, and the clergy used the Crucifixion story as the moment to illustrate the demonic nature of "the Jew." The rituals of Holy Week provided the clergy with a clear narrative to fix in the minds of the faithful the enormous crime that the Jews had committed against Jesus.[66]

The symbolism of Christianity is extremely powerful; its emergence and predominance as a major religious and political force make this obvious. It is the narrative quality of Christianity that is so powerful, a symbolic progression that makes its way into popular thought from sermons and church services to the Passion Plays, stations of the cross, Carnival parodies and, in Haiti, Vodou myth and Rara theater. These popular, dramatic manifestations demonstrate that Christianity is not just a theological system: "It is also a structure of the imagination; its most striking feature, by which it has gained its hold on a great portion of humanity, is its narrative."[67]

Bruce Lincoln admonishes us that examining "world religions without considering the way they came to enjoy that status is a serious methodological and moral error. In contrast, to probe that history is to understand how a relatively small set of religions have expanded their territory and power at the expense of others."[68] In Haiti, the site of the first American colony, the African survivors of plantation slavery have inherited the anti-Judaism of medieval Europe. At the same time the same Africans and their Creole descendants became targets of a brutal campaign of cultural violence that used anti-Judaism as its blueprint. Today, layers upon layers of historical symbolic residue swirl around together, as "Jews" are both saints and ancestors, pigs and Judases. They can be at once ancient wisdom and global capital, but are always a mystical and exotic power in the Haitian imagination.

Notes

1. This work is indebted to Phyllis Mack and her faculty research seminar at the Rutgers Center for Historical Analysis, where I began this essay as a postdoctoral fellow, and to Albert J. Raboteau and the Northeastern Seminar on Black Religions at Princeton University. I also wish to thank my research partners in Haitian fieldwork: Chantal Regnault, photographer extraordinaire; my assistant, Phenel Colastin; and our driver, Blanc Bazle. Thanks also to Father Antoine Adrien and Holly Nicolas, and to Leslie Desmangles, Joel Dreyfuss, Henry Gold-schmidt, Leon-Francois Hoffman, Glenn Ingram, Alan Nathanson, Judith Weisenfeld, and Jeremy Zwelling. I am also most grateful to Deborah Dash Moore and the Young Scholars Program in American Religion and Betsy Trauble and the Wesleyan University Center for the Humanities, and wish that space permitted me to name the participants in these seminars who generously helped me think this story through.
2. Bartolome de las Casas, *History of the Indies* (New York: Harper and Row, 1971). Cited in Catherine Keller, "The Breast, the Apocalypse, and the Colonial Journey," in *The Year 2000: Essays on the End*, ed. Charles B. Strozier and Michael Flynn (New York: New York University Press, 1997), 42–58. Haiti has various name changes: The Amerindian "Aiyti-Kiskeya" was changed by Colon to "Hispaniola," "Little Spain." Later, in 1697, the French named their colony Saint-Domingue, and in 1804, newly independent slaves and people of color returned the land to its original name of Haiti.
3. These themes are taken up in my dissertation; see Elizabeth A. McAlister, " 'Men Moun Yo,' 'Here Are the People': Rara Festivals and Transnational Popular Culture in Haiti and New York City," PhD dissertation, Yale University, 1995.
4. Ronald Sanders, *Lost Tribes and Promised Lands: The Origins of American Racism* (1978, repr., New York: Harper Perennial, 1992), 90.
5. Cecil Roth, *A History of the Marranos* (Philadalphia: Jewish Publication Society of America, 1932). *Anusim*, Hebrew for "forced ones," has now replaced the English "Crypto-Jews" or the Spanish *conversos*, or "converted Jews," and the more derrogatory *marranos*, or "swine," in Jewish studies literature. See also David M. Gitlitz, *Secrecy and Deceit: The Religion of the Crypto-Jews* (Philadelphia: Jewish Publication Society, 1996).
6. Christendom also links the Jews, the devil, and the peoples of the First Nations in New Spain and the United States. See Fernando Cervantes, *The Devil in the New World: The Impact of Diabolism in New Spain* (New Haven, CT: Yale University Press, 1994).
7. Michel-Rolph Trouillot, *Haiti, State Against Nation: The Origins and Legacy of Duvalierism* (New York: Monthly Review Press, 1990), 81.
8. World Bank, "Memorandum on the Haitian Economy. May 13. Latin American and Caribbean Regional Office, 1981." Cited in Linda Basch, Nina Glick Schiller, and Cristina Szanton Blanc, *Nations Unbound: Transnational Projects, Postcolonial Predicaments and Deterritorialized Nation-States* (Langhorne, PA: Gordon and Breach, 1994), 159.
9. Basch, Schiller and Blanc, *Nations Unbound*, 161.
10. See, e.g., Joan Dayan, *Haiti, History, and the Gods* (Berkley: University of California Press, 1995); Leslie Desmangles, *The Faces of the Gods: Vodou and Roman Catholicism in Haiti* (Chapel Hill: University of North Carolina Press, 1992); LeGrace Benson, "Some Breton

and Muslim Antecedents of Vodou Drapo," in *Sacred and Ceremonial Textiles* (Chicago: Textile Society of America, 1996), 68–75.

11. Karen McCarthy Brown, *Mama Lola: A Vodou Priestess in Brooklyn* (Berkeley: University of California Press, 1991).
12. See Serge Larose, "The Meaning of Africa in Haitian Voodoo," in *Symbols and Sentiments: Cross-Cultural Studies in Symbolism*, ed. I. M. Lewis (London: Academic Press, 1977), 85–116.
13. In Vodou, the soul is made up of different overlapping parts. The *zonbi* is a part of the soul that lingers near the grave. It retains the personality of the living person and can be captured and manipulated. This *zonbi*, the *zonbi astral*, is not connected to a body and is different from the living-dead figures in the popular Haitian and U.S. imagination, so often portrayed by Hollywood. For an elaboration on the uses and construction of *zonbi astral*, see Elizabeth McAlister, "A Sorcerer's Bottle: The Art of Magic in Haiti," in *Sacred Arts of Haitian Vodou*, ed. Donald J. Cosentino (Los Angeles: UCLA Fowler Museum, 1995), 304–321.
14. This is the time of Jesus' death noted in Scripture.
15. Thérèse Roumer, interview, Petionville, February 16, 1993.
16. George Fouron, personal communication, New Haven, CT, November 1997.
17. Thérèse Roumer, interview, Petionville, February 16, 1993.
18. W. B. Seabrook, *The Magic Island* (New York: Literary Guild of America, 1929), 270–272.
19. See, e.g., Muriel Thayer Painter, Edward H. Spicer, and Wilma Kaemlein, eds., *With Good Heart: Yaqui Beliefs and Ceremonies in Pascua Village* (Tuscon: University of Arizona Press, 1986); and James S. Griffith, *Beliefs and Holy Places: A Spiritual Geography of the Primeria Alta* (Tuscon: University of Arizona Press, 1992), 95.
20. See Marilyn Ekdahl Ravicz, *Early Colonial Religious Drama in Mexico: From Tzompantli to Golgotha* (Washington, DC: Catholic University of America Press, 1970); and Richard C. Trexler, "We Think, They Act: Clerical Readings of Missionary Theatre in 16th Century New Spain," in *Understanding Popular Culture: Europe from the Middle Ages to the Nineteenth Century*, ed. Steven L. Kaplan (New York: Mouton Publishers, 1984), 189–227.
21. In *Judas Iscariot and the Myth of Jewish Evil* (New York: Free Press, 1992), Hyam Maccoby points out the consistent use of Judas by Christian myth as a symbol for all Jews. "Of all Jesus' twelve disciples, the one whom the Gospel story singles out as traitor bears the name of the Jewish people."
22. Sander Gilman, *The Jew's Body* (New York: Routledge, 1991), 18.
23. Joshua Trachtenberg, *The Devil and the Jews: the Medieval Conception of the Jew and Its Relation to Modern Antisemitism* (1943; repr., New York: Harper Torchbooks, 1966), 20.
24. Trachtenberg, *Devil and Jews*, 20.
25. Painter, Spicer, and Kaemlein, *With Good Heart*, 352.
26. Trachtenberg, *Devil and Jews*, 22.
27. Trexler, "We Think, They Act."
28. For a discussion of the conflation of British, Protestant, and Civilized into one identity against Native American "heathens," see James Axtell, *The Invasion Within: The Contest of Cultures in Colonial North America* (New York: Oxford University Press, 1996).
29. See Roth, *History of the Marranos*. The anti-Judaism taught by the Catholic clergy in Haiti bears the characteristics of a classically premodern Jew-hatred centering on the betrayal of Judas. In this logic, Jews are primarily polluters and traitors; there is little reference to the modern anti-Semitic tropes of a Jewish conspiracy of exploitation hinging on issues of capital or usury. See the chapter "From Anti-Judaism to anti-semitism" in Gavin I. Langmuir,

History, Religion and Antisemitism (Berkeley: University of California Press, 1990), 275–305.

30. On the Inquisition and the Jews in Mexico, see Seymour B. Liebman, *The Jews in New Spain: Faith, Flame and the Inquisition* (Coral Gables, FL: University of Miami Press, 1970).
31. Trachtenberg, *Devil and Jews*, 26.
32. See Marc Shell, "Marranos (Pigs), or from Coexistence to Toleration," *Critical Inquiry* 11, no. 2 (Winter 1991): 306–335.
33. Carolyn E. Fick, *The Making of Haiti: The Saint Domingue Revolution from Below* (Knoxville: University of Tennessee Press, 1990), 278.
34. George Breathett, *The Catholic Church in Haiti (1704–1785): Selected Letters, Memoires and Documents* (Salisbury, NC: Documentation Publications, 1982), 4.
35. Anne Grene, *The Catholic Church in Haiti: Political and Social Change* (East Lansing: Michigan State University Press, 1993), 76.
36. G. Debien, "La Christianisation des esclaves des Antilles francaises aux XVIIe et XVIIIe siecles," *Revue d'histoire de l'Amerique francaise*, 21 (1967): 99–11.
37. Ephesians 6:5.
38. See Joshua, *Devil and Jews.*
39. Dayan, *Haiti, History, and the Gods*, 252.
40. Ibid., 253.
41. Moreau de Saint Méry [1797] 1958, 1:55. Cited in Desmangles, *Faces of The Gods*, 27.
42. Dayan, *Haiti, History, and the Gods*, 252.
43. Ibid, 253.
44. Cited in Laënnec Hurbon, *Dieu dans le Vaudou Haïtien* (Port-au-Prince: Editions Deschamps, 1987), 21.
45. David H. Brown, "Garden in the Machine: Afro-Cuban Sacred Art and Performance in Urban New Jersey and New York," PhD dissertation, Yale University, 1989, 16.
46. Peter Stallybrass and Allon White, *The Politics and Poetics of Transgression* (Ithaca, NY: Cornell University Press, 1986), 3.
47. I witnessed such a Passion Play by African American Catholics at Saint Ann's shrine in New Orleans, for example, in 1998.
48. The band Ya Seizi can be heard playing in Papa Dieupe's compound on track 19a of the recording compiled by this author: "Rhythms of Rapture: Sacred Musics of Haitian Vodou," Smithsonian/Folkways Recording SF 40464, 1995.
49. Papa Mondy Jean, interview, Port-au-Prince, April 1992.
50. A few Rara presidents told me that there was a game, a noisemaker, that the Jews held in their hands and spun at the Crucifixion. This made a noise that came to be called "Rara." Papa Telemarque, interview, Darbonne, Léogane, March 6, 1993. One notices the possible connection with the noisemakers of Purim. This connection seems obscure; at any rate, it is yet impossible to locate.
51. Rara costumes are elaborately sequined in parts of Haiti. Siméon, interview, Bel Air, Port-au-Prince, July 30, 1993.
52. Siméon, interview, Bel Air, Port-au-Prince, March 20, 1993.
53. Siméon, interview, Bel Air, Port-au-Prince, July 30, 1993.
54. Laënnec Hurbon, *Culture et Dictature en Haiti: 1'imaginaire sous contrôle* (Paris: L'Harmattan, 1979), 43.
55. Alfred Métraux, *Voodoo in Haiti* (1959) (New York: Schocken Books, 1972), 326.
56. Papa Dieupe, interview, Artibonite, Easter Sunday 1993.
57. A similar symbolics works in Afro-Cuban religion, Lukumi. Unbaptized ritual objects and "working" charms are called *judeo*, "Jewish."

58. Mayard, interview, Rara Mande Gran Moun, Léogane, March 20, 1993. David H. Brown reports an interesting parallel in the Kongo-derived Palo Monte practices in Cuba. As he constructs a *prenda*, a "working" object, on Good Friday, a Mayombero comments to Brown, "On the day of the week, the week of the year when they are quiet—Good Friday—we are doing our thing." Says Brown, "As spiritual opposites of Christ and the Saints of Olofi and the orichas, they are 'driving nails' on the day of the Crucifixion." Brown, "Garden in the Machine," 375.
59. See Hurbon, *Culture et Dictature en Haiti*, 133.
60. Liebman, *Jews in New Spain*, 31.
61. Clarence Haring, *The Spanish Empire in America*, 29. Cited in ibid., 42.
62. Liebman, *Jews in New Spain*, 42.
63. Zvi Loker, "Were There Jewish Communities in Saint Domingue (Haiti)?" *Jewish Social Studies*, 44, no. 2 (Spring 1983): 135–146.
64. Ibid., 137.
65. The residence of the Israeli ambassador, at the time of this writing, is at the center of Jewish activities in the country. He invites any practicing Jews and Jewish visitors to Haiti to holiday seders at his home.
66. Bernard Glassman, *Anti-Semitic Stereotypes Without Jews: Images of the Jews in England 1290–1700* (Detroit: Wayne State Press, 1975), 25.
67. Maccoby, *Judas Iscariot*, 1.
68. Bruce Lincoln, "Religious Imperialism and Its Victims: Resisting the Erasure of Those Who Resist," conference paper, UC Davis, March 1996.

Chapter 7

Resisting Freedom: Cultural Factors in Democracy—The Case for Haiti

Patrick Bellegarde-Smith

The trees fall from time to time, but the voice of the forest never loses its power. Life begins.

—*Jacques Alexis,* Les Arbres Musiciens *(Paris, 1957)*

Haitian societal development seems to be filled with paradoxes, contradictions, and enigmas. The process begun with the "encounter" of 1492 was profoundly corrupting in terms of cultural deflections to both the Old World and the New World and to established patterns of interaction everywhere between the powerful and the weak that are still evident today. Haitian history has been marked by struggle and by continuous acts of rebellion that, in more quiescent times, have become passive resistance known by the Haitian label, *marronage*.

The Haitian Revolution of 1791, the most complete social revolution in the hemisphere until Mexico in 1910 and Cuba in 1959, was formed by two major distinct movements that coalesced in an unsteady alliance but never fully merged. The first movement, that of the enslaved, was the heir to all earlier slave rebellions, and was consistent with slaves' opposition to slavery and all forms of oppression. The second movement was that of the plantocracy of color. Using a combination of tactics and approaches such as warfare, guerrilla activities, and poisonings, this group terrorized the plantocracy. The leaders were often Vodou priests. Boukman Dutty and a female priest are said to have officiated at a Vodou ceremony at Bois Caïman—doubtless one of many such ceremonies—inaugurating the insurrection of 1791. Makandal, an African-born Muslim with a reputation of being a *bòkò*, led the 1757 revolt.[1] Pierre Dominique Toussaint Louverture, who later ruled, was also thought to be a bòkò. Makandal's sacred mission, like that of Zumbi of Palmares, was to rid the colony of its white inhabitants and to create an African kingdom in Saint-Domingue.

The social philosopher Dantès Bellegarde (1877–1966), who was most unsympathetic to Vodou, wrote the following about Makandal: "The certainty [that he had survived execution by the French] played an important role in organizing later uprisings. It maintained the trust of slaves who had found in the Vodou religion a particularly strong ferment to exalt their energies, since Vodou . . . had become less a religion than a political movement, a kind of 'black *Carbonarisme*' whose objective was white extermination and black liberation."[2]

Writing later about Toussaint Louverture, Bellegarde stated: "Some have doubted the sincerity of the Catholic faith of Toussaint without offering a single fact to justify such a doubt." He continued in an apparent contradiction, "he knew from personal experience that the vodun ceremonies were only a pretext for political reunions where, in the secretive 'houmforts' [temples] and in the exaltation of ritual dancing, plots were being hatched against the authorities or, again, attacks against property; [Toussaint] had become the 'authority' [as governor] and had made himself defender of property, which article 13 of the Constitution had declared 'sacred and unassailable.' "[3] Under capitalism, property is sacred and freedom of cultural expression can be curtailed in its defense. The slaves qua peasants had been forced to remain on plantations against their will. They revolted, as they would again and again *after* Haitian independence, for which they had fought. Most of the 100,000 deaths during the war for independence were from their ranks.

Toussaint Louverture, as well as all of his successors at the helm of the Haitian state, realized Vodou's potential for challenging the power and authority of the state and the "mainstream" vision of society. Toussaint Louverture was thought by his followers to be a believer in Vodou, reflecting the phrase that "Haiti is 80 percent Catholic, but 100 percent Vodou." This emphasizes the view that it need not be "practiced" to have a valid hold on most minds as a state of mind, a spiritual system, and a world view.[4]

By the time Toussaint Louverture came to power, an awesome change in the nature of the struggle had already begun. The leaders of the first uprisings had died in battle, been assassinated, or been co-opted by newcomers who were themselves Westernizers. Plymouth, Jean Biassou, Lamour Dérance, Jean-François Papillon, Romaine la Prophétesse, and, of course, Boukman Dutty, and countless other men and women were "dismissed" in a later phase of the struggle for independence. Westernizers in power have dismissed competing historical explanations of various events.

The majoritarian vision in Haiti, as elsewhere, was established largely on the collective experience of enslavement that, at its onset, had solely an economic purpose: to exploit the natural resources and human capital cheaply for the benefit of the metropolitan elites. Indeed, the colonial state—with intentional as well as unintentional results—set social policies to sustain economic exploitation in the plantation system by converting the enslaved population to Christianity and by assimilating a few to French cultural norms in order to enslave better. Some slaves saw the benefits of such a course of action. The majoritarian vision also developed from the peasants' shared language (Kreyòl) and religion (Vodou) which helped provide them with an ideological superstructure in form *and* content.[5]

Significantly, language and religion were inventions of the Haitian peasants that permitted them to face the power deficit between master and slave, oppressor and oppressed, and a dominant Europe and a subaltern Africa. In both Haitian language and Haitian religion, one finds the necessary compromises and subterfuges that the weak use to survive. They also reflect a "hybridity" built largely on deception, later referred to by some scholars as "Creole culture." But synthesis does not mean syncretism. Neither implies synergy, but sometimes its opposite.[6]

Kreyòl and Vodou are, in the final analysis, organized responses to oppression, created *from* African foundations. Kreyòl appears to have grammatical and syntactical roots in West Africa, notably in the Kwa language group, into which foreign words (such as French) were easily incorporated out of the necessity to communicate with those in power.[7] Kreyòl is conceptually rich, having borrowed from many sources to reflect new artifacts. At no time was Kreyòl "animal talk," "baby talk," "*petit nègre*," "*un français remis en enfance*," or "bad French," as long believed by many scholars. At no time was Vodou "*un Catholicisme maldigéré*," although it may have been *indigeste*. Both emanated from West African societies remarkable for their overall cultural similarities, although there was certain preeminence achieved over time by a core of Fon, Ewe, Yoruba, and related peoples.[8]

The inclusion of non-African forms into the Haitian language and religion became a form of adaptation rendered necessary by circumstances. Kreyòl became, at once, a "contact" language and a "counterlanguage," the latter with its studied ambiguity, irony, satire, and wit, and its ability to *voye pwent*.[9] In one sense, it defeated the purpose of separating Africans from members of their ethnic group. As a contact language, it became the lingua franca in Saint-Domingue between French and Africans and between African and African. Today, all Haitians speak Kreyòl; less than 10 percent of the population is fluent in French, the other official language. As the national language understood by all, Kreyòl served as the language of widest communication throughout Haiti, particularly in the absence of widespread literacy.[10]

Born in the Americas, Vodou responded to the spiritual and temporal needs such as in health care and psychology of the majority population by amalgamating rituals from west, west central, and south central Africa. Here again, Dahomean and Yoruba elements predominated, with significant contributions from Kongo cultures and civilization.[11] A strong family resemblance exists within the religious systems, moral ideals, and social organizational structures of most African cultures, particularly as concerns ancestral veneration and the cosmic, esoteric, or occult interaction between worlds and universes. The significance of Vodou in the Haitian struggle for independence, in the slave uprisings that preceded it, and in subsequent peasant rebellions, cannot be overstated, nor its importance in most art forms such as music, dance, the plastic arts, and oral literature, as befits a system that is also the culture's repository for social thought, organization, and ideology.

The central role of religion in human affairs is fairly well understood for its explicatory value in all societies. Less a religion than a spiritual system and discipline, Vodou extends "naturally" and easily, permeating all systems, structures, and institutions in the large and subtle ways in which most Haitians view their world and all worlds, even as they migrate.[12] Indeed, Haitians need not be observant, initiated, nor believers to partake of that shared national world view fostered through Vodou.[13] As with Judaism in Israel and Shinto in Japan, Vodou helped define the Haitian nationality and ethos. Roman Catholicism, in its French variant, remained the official religion until 1987 and defined the state, the powers that structured it, and the elites who controlled it. Individuals who have decried Vodou or who have been embarrassed by it have hoped to attain a "higher standard," an interesting point of colonial psychology.[14] The ostensible rejection of Vodou, or any other African cultural marker, has been absurdly common among black populations throughout the Western Hemisphere, where physical oppression has been accompanied by the requisite imposition of European cultural norms, resulting in the marginalization of these groups. They have, in turn, internalized their apparent inferiority and have been given the status of a minority (not necessarily defined in demographic terms). Colonialism engaged all human faculties and has been a factor in all social science.[15] Eurocentric anthropologists, psychologists, and economists, typically anchored in their own civilization, have found it necessary thus far to divide the world between "primitive mentalities" and modern man, and in perfectly rational arguments support directly policies that subjugate most areas of the world.[16]

Power That Defines: Class, "Modeling," and Dominant Paradigms

The leaders who hijacked the twenty-one American republics in the nineteenth century, leading them to formal independence, had come from similar (if not identical) social backgrounds as *criollos* (in Spanish America) and colonists (in the United States). Both groups were descended from Europeans who settled these colonies after they had been "discovered" by their forebears, who had come as conquerors.

The Haitian struggle for independence, formally lasting thirteen years, led to the creation of the first independent Latin American state in 1804. It was, besides, the first "modern" state of African origin on the Western model, followed by Liberia, whose elites were established along identical premises.

The Haitian war of independence came about as a result of an amalgam of diverse movements and conflicts. The first movement arose on the part of the enslaved majority—about 90 percent—as a search for freedom and liberation, defined by them as cultural autonomy and the power of political initiative. The second movement erupted from among the middle groups whose existence came from the social order fostered by the encounter between the West and Africa in colonial Saint-Domingue, as elsewhere. Indeed, between the white apex and the black base of the societal triangle, there existed a level intermediate in status and color, which aspired to be closer to God.[17] Class interests dictated that the affranchis, as the free people were called, would lend their support to whichever group could buttress its insecure social position. Middle classes in all cultures have often felt inherently insecure. In perhaps 99 percent of the cases, they have exercised the option of lending their support in favor of the powerful upper classes. The French Revolution of 1789, *not* the endemic slave uprisings, had unleashed the affranchis' natural reserve, as they wished to remove the petty racial vexations of colonial social apartheid and to become *citoyens actifs* as the revolution had promised.

The affranchis had achieved a significant economic position despite French racism: One-fourth to one-third of the plantations and one-fourth of the slaves belonged to them. The white plantocracy had misgivings about the racial implications of the French Revolution with respect to colonial society. Consequently, many in the white plantocracy were royalists who had sided with France's arch rival, the United Kingdom. Spanish settlers in the Americas had had the same qualms about a liberalizing Spain (under Napoleonic rule), a concern that had led to Mexican independence under Augustín Iturbide, whose group came to power partly to forestall a social revolution from among the (Indian and mestizo) lower class in the early nineteenth century.[18] U.S. intervention in the Spanish-American-Cuban War had a similar genesis. The Haitian affranchis had achieved a status similar to that of the Spanish-American criollos and the North American colonists, with similar conflicts arising in Saint-Domingue as had arisen between criollos and *peninsulares*, the American colonists and the British. In Haiti that dispute carried an additional burden, that of biological "race." The white plantocracy opposed the granting of social and political equality to brown and black planters. What had been a family dispute among putative members of a potential same bourgeoisie would lead to a "racial" war because of the intransigence of white Saint-Dominguois and French racism.

Colonial societies such as Haiti, in which a segment of the population is often formed from the union between ruler and ruled and master and slave from disparate cultures, frequently have as a salient element of their psychology the self-hatred and sense of violation resulting from the cultural and emotional rape that took place. This sense of violation and distrust accounts for tensions in the existing relations between the governed and the social institutions of the country. By forcibly depriving people of their cultural heritage, colonialism sought willfully to structure and organize all social alliances, thereby instituting its domination over the norms and values of the colonized.[19] More often than not this domination resulted in feelings of inferiority among those who had been colonized, in part the very definition of a racial minority. The vanquished often exhibit a form of idolatry vis-à-vis the conqueror. This sort of psychological violence inherent in all colonization—and in religious conversion—together with most colonial administrative practices, goes a long way toward explaining the spirit and the reality of authoritarianism found in neocolonial societies. Often unable to transcend the conditions of their birth, Haitian elites are also victims of the clash between the Western old and a new non-Western world order. Furthermore, in contradistinction with the rigid racialist position of Anglo-Saxon cultures—de jure policies as well as de facto—cultural assimilation was promoted by the Latin powers. Anglo-Saxon societies openly excluded blacks; Latin societies, in theory, included them, but only according to a racialist argument that blacks could be salvaged, despite their having nothing to contribute culturally. Christianity everywhere adopted the Latin version of racism by actively seeking converts from among black populations. Under such odds, what was to be the rational response of Haitian elites?

The grand alliance between slaves and affranchis that culminated in Haitian independence fourteen years later, in fact, was a cynical and calculated effort to co-opt the slaves' fighting power for the affranchis' own purposes.[20] That alliance was, nonetheless, the necessary condition for independence, although not for freedom.

The assessment of the philosopher Louis-Joseph Janvier (1855–1911) is particularly useful in this context. Commenting on Haiti's continued authoritarian tradition, he wrote at the turn of the century that "political freedom is an inferior good [compared to] national independence."[21] Ribeiro, the Brazilian scholar, synthesized the conditions left by colonialism, stating, "Each people, even each human being, was affected by and caught up in the European economic system or the ideals of wealth, power, justice, or health inspired by it."[22] This legacy also explained Emperor Jean-Jacques Dessalines' cry of anguish when former affranchis tried to appropriate all property that belonged to their white fathers (not to their black mothers): "What of the poor blacks whose fathers are in Africa, won't they inherit anything?" In the larger context, however, it meant slavish acceptance of liberal ideology as it sustained capitalism, which was seen as quintessentially "modern." In early nineteenth-century Haiti, French political influences, British social thought, and the example of the North American rebellion of 1776 aided in the formulation of Westernized ideologies that were viewed as rationalistic and the sine qua non of modern existence.[23]

The ascendancy of the former affranchis occurred shortly after independence. The consolidation of elite hegemony, starting in the 1820s under the long reign of President Jean-Pierre Boyer, led to the first major postindependence peasant uprising, in 1843. The Piquets, the "suffering army" as the insurgents called themselves, demanded land distribution, the dispossession of the rich, and "black" (peasant) political control. Beaubrun Ardouin, a historian (1796–1865), had claimed the Caribbean as a whole for the *mulâtres* (a new name for the former affranchis) who had nowhere else to go. Simón Bolivar had made a similar argument for Spanish America.

Despite the "revolutionary" message, there was the continuation of structures, the form and the spirit of colonial constructs adopted from the slave master. The historian Lespinasse (1811–1863) added to the analysis that the spirit of freedom was a mulâtre gift to the blacks—that the slaves had acquired a taste of independence from mulattos. But all must now unite for *raison d'état*.[24] These historians had rendered in a fairly sophisticated way—the first time this had been done by Westernized blacks anywhere in the world—even earlier analyses published by Haitian intellectuals. In 1817, for instance, Valentin Pompée de Vastey had defended the European colonization of Africa almost a century before it was fully under way.

Whether based on pre-Enlightenment ideologies, or liberalism, conservatism, positivism, or, much later, Marxism and modernization theories, all were ideas expressed by the small coterie of Haitian intellectuals that hoped to help define Haiti to the rest of the world, institute its policies, and define its raison d'être. All of these ideas came from a Western civilization that claimed universality, but in fact remained alien to most Haitians. Once translated into state policy, these ideas were never so broad as to incorporate broad segments of the polity into *res publica*—neither the very large peasantry, the small urban working class, or at its outset, the emerging middle classes.

The desire of one group to maintain power over others occurred through a brutal econo-political system instituted from the ideological base of a political culture dominated by an upper class, and not necessarily from a national political culture. Elites were fearful of relinquishing pieces of what was a limited pie, which, starting in the 1880s, was also additionally shared by expatriate European, North American, and Levantine elites.[25] Individuals and groups vied for preeminence within a restricted and restrictive social environment, in which perhaps 5 percent of the population counted.[26]

Among the elites, men (and a few women) argued in favor of the French language and of Roman Catholicism as the pillars upon which to erect Haitian culture; some mistook the dream for the reality. The obstacles to achieving this on a widespread basis eventually proved insurmountable. Most Haitians knew that Kreyòl and Vodou were the *poto mitan* (the center posts) of Haitian culture, the language they spoke and the gods they worshipped. The dichotomies in Haiti near the turn of the century between European and African, elites and peasantry, French and Kreyòl, and Catholicism and Vodou were vital echoes of earlier dichotomies between affranchi and slave, brown and black, creole and *bossale*, and civilized and primitive. The dichotomies informed a vision, a world view, and a mentality in which reality, to a large extent, was composed of opposites, and was viewed as such. Haitian society was in fact ordered according to a non-Western view in which knowledge of French and Kreyòl was part of a continuum, in which the majority worshipped the *Lou* (deities) while still going to church, and where shades of brown and black occurred within each extended family. In the final analysis, the national period reflected a continuation of social policies set by France during colonization that were, in a paradox more apparent than real, easier to pursue once the whites had been forcibly removed from the national scene.

The first American occupation of Haiti (1915–1934), brutal at all levels, was a cultural shock as well.[27] In view of the proud and bellicose racism of the United States as it applied to the whole of Latin America, Haiti—*la France noire* of Jules Michelet—was believed by the United States to be inoperative. Early on, Simón Bolivar had stated that the United States, "so enamored of its own liberties, is already less fond of the liberties of others. Quite the contrary: it has made of this liberty an instrument for causing other people misery."[28] U.S. economic interests in its private sector, far more than strategic interests, had been a primary cause of the occupation.

In the shock of the occupation, emerging middle class intellectuals throughout Haiti rose against the cultural status quo, defending African norms and values adhered to by the majority of Haitians, despite knowing little about them. These efforts did not preclude widespread persecution of Vodou, however, nor did it signify an acceptance of Kreyòl. Hurt to the very core of their souls by an American army that had penetrated Haitian psychic defenses, intellectuals rebelled against the vision expressed by Bellegarde that Haiti was "an intellectual province of France." Later, in the heat of the culture wars he would add, "What would happen to a Dahomean islet in the heart of the Americas?"[29] The *Indigénisme* movement spread nonetheless, inching closer—but not too close—to a peasantry lauded in the abstract.[30] That peasantry had taken the brunt of the U.S. invasion and had responded to that invasion with an armed resistance movement. It appeared to the peasantry that the whites were reinstituting slavery.

Bellegarde's nemesis was his personal friend, Jean Price-Mars (1875–1969), whose works were well placed to "start" the worldwide movement of négritude, yet another literary movement of blacks who had been assimilated into French culture, and who had realized faintly what had been lost. The more radical members of these groups moved toward Marxism. But both Marxism and négritude were still solidly anchored in a Western ethos—an anti-Western Westernism, if you will—similar to Sartre's description of négritude as racist antiracism.[31] These Haitians retained an etic viewpoint; in the context of Haitian upper and middle classes and education, an emic quality remains elusive.

Not so paradoxically, the U.S. occupation shored up the Haitian upper classes, the best guarantors of U.S. hegemonic economic interests in the country, against the rising popular tide. It had been increasingly difficult to quell popular discontent against a status quo established in 1806. A series of ephemeral governments at key points in Haitian history had revealed the disarray of a ruling group faced by pressures from below. At times, only foreign intervention seemed able to maintain political stability.

By 1847, the system had already undergone tremendous stress, largely from the Piquets' uprising. The country had four presidents between 1843 and 1847. As the malaise continued to grow, and rural uprisings maintained pressure on Port-au-Prince—the Caco rebellions might be said to have lasted between 1860 and 1929—six ephemeral governments succeeded one another between 1911 and 1915. Later, additional pressures from the newly conscious middle classes, as distinct from the peasantry, led to five governments between 1956 and 1957. The crises became increasingly severe. After the fall of President Jean-Claude Duvalier in February 1986, five provisional governments took power between 1986 and 1991, and a prolonged military dictatorship under General Raoul Cédras took over, with three distinct administrations between 1991 and 1994.[32]

In each case, the crisis was resolved temporarily—by President and later Emperor Faustin Soulouque in 1847, the American occupation in 1915, the "selection" of President François Duvalier in 1957 with U.S. support, and the election of President Jean-Bertrand Aristide in 1991 and his restoration in 1994, with U.S. support. The United States proved to be a crucial player in the transition between upper class and middle class rule in 1957 and between middle class and "popular" rule in 1991, while its hegemonic interests remained intact or actually increased. A symmetry was being established: Desperate living circumstances would lead to revolutionary conditions, which, in turn, would lead to repression. Each response ultimately led to increased repression.

As if *le Peuple* Mattered: Democracy in Popular Culture

Demographics and class structure, as well as popular culture dictate that the impact of social movements upon the body politic and political culture be considered seriously. The objective had been to stymie popular participation in politics. Frequent declarations about the inherent lack of a democratic tradition in Haiti are ahistorical and are anchored in a Western social science tradition that rejects the reality of class, while itself remaining culture specific.[33] More significant in furthering an understanding of Haiti's political culture might be the study of the ideological superstructure ensconced in a metaphysical world view elaborated by Haitians. Palmares and other similar experiments in indigenous autonomy were destined to fail when at odds with the economic interests of the powerful. Indeed, the cry "*que deviendrait un îlot dahoméen* . . . ," however ill intentioned and racist, acknowledged that Haiti could be recolonized *because* it was barbaric, as was Africa in the 1940s. The statement did not, however, question the basis for that assertion. The elites' best intentions notwithstanding, they did not preclude neocolonization starting in the 1880s or self-righteous efforts by the international Western community to civilize Haiti.[34] But belles lettres and belles manières did not suffice in saving the country from interference and intervention.

Middle class political control (as distinct from economic power) had incited great hopes in the 1940s. The middle classes did not bring with them the psychological baggage or the reservations of the previous ruling group regarding language and religion. And it claimed to represent *le peuple.* Cultural nationalism transmogrified into political nationalism, and were both a part of an anticolonial arsenal in which the master, nonetheless, remains the model, the *patron* (blueprint or boss) for one's action. These developments indicated that after two centuries of independence, Haiti and the still-colonized Caribbean countries were not far apart politically or culturally. The middle class faced social apartheid and petty humiliations of the kind suffered earlier by the affranchis, whose descendants and intellectual heirs they were now fighting. That middle class had been formed by the economic space created by beneficial worldwide conditions, the political stability provided by the U.S. occupation (that also discriminated against it on the basis of color), and the educational opportunities established by the governments of Philippe-Sudre Dartiguenave and Louis Borno. In fact, the middle class represented a new element in the equation in the *rapport de forces* that failed to live up to the expectations of many.

Although opposed to the first middle class government of President Dumarsais Estimé (1946–1950), the United States supported the second, that of François Duvalier. The United States seemed fairly comfortable with a racial rather than a class analysis, since it could and did blunt class differences while augmenting the "political class" and giving the appearance of expanding democracy. The United States had followed a similar course in Cuba and the Dominican Republic with success. As was later proven within the confines of the United States proper during the Civil Rights movement, an obsession with the social construct "race" (as a form of nationalism), is more easily a part of a conservative and traditionalist agenda than not.[35] The old guard, nonetheless, had viewed the replacement of cherished cultural positions (Haiti as Christian, French, and a part of the Western world), as challenging a national identity, disguising class interests.

The issue of a national identity (hence, personal identity) had a profound impact and implications in all other areas.[36] But for two centuries, the debate took place in a rarefied atmosphere in which the majority of Haitians had no voice. Despite real cultural dislocation, as the intelligentsia peregrinated from Pan-Americanism, to Pan-Caribbeanism, to Pan-Africanism, indigénisme, negritude, and now to Antillanité—all a search for a middle ground and a sustainable identity—the people seemed remarkably unconcerned. The uprisings of yesteryear seem to have been replaced by a form of passive resistance, a *marronage*, reflected in religion, language, and fairly uncontrollable institutions and activities that set the province apart from the

"République de Port-au-Prince," the countryside from the urban center when the center's centripetal force was increased purposefully during the occupation.[37] The "*gouvenman lannuit*" was harder to sustain, as the reach of the central government became more realizable.[38] And the widespread and generalized movements that led to the overthrow of the dictatorship in 1986 had come from all segments of the population, including the Catholic Church, the *hounfò* (Vodou temples), Duvalierists (who attempted Duvalierism sans Duvalier), and the army. But because of irreconcilable differences, the grand alliances broke as soon as they coalesced. The popular movement remained unimpressed by political parties (a sine qua non of Western definitions of democracy) until Lavalas came upon the national scene.

But where there is intrusion there is resistance. Only in the area of culture can an oppressed population find a playing field. That resistance was exemplified and illustrated in a culture that was often countercultural and would become the dominant culture if there were a change in the socioeconomic system of the sort that might have been sustained if the Haitian Revolution had survived the assassination of Jean-Jacques Dessalines in October 1806. President Jean-Bertrand Aristide was conscious of both symbol and reality when he took the oath of office in Kreyòl, and received the presidential sash from a *manbo* (female Vodou priest) in February 1991, with symbolic forms of greeting that resonated deeply within the population. His deliberate use of Kreyòl indicated a break and discontinuity in cultural patterns that was strangely symbolic of the synthesis in Vodou and dealt blows to both the elite language (true language, French) and the elite religion (true faith, Catholicism).[39] Subsequently, in 1995 Aristide further evened the playing field between Vodou and Christianity. When he flung open the gates of the presidential mansion to serve food to beggars personally, he simultaneously transformed a white house built on the model of the Petit Trianon into a national palace.

In all fairness to France, the spread of the French language, which was essential in establishing the commonweal, was incomplete by the time of Haitian independence. France was not fluent in French. The French Revolution and the Napoleonic Wars were the context in which that language spread inside France, creating the modern nation state and nationalism as we now understand it. Haiti was linguistically unified before France. The adoption of French (rather than English, which Henri Christophe might have wished) rendered that language of differentiation and, in the process, made Kreyòl the language of countercultural resistance. The manichean dichotomy, which is not found in the Vodou religion, need not have occurred: A linguistic continuum has always existed, illustrated by the expressions *parler pointu* and *parler plat.*[40] Language has "color" and becomes a marker for limiting access to power.

Carew, the Guyanese novelist, said that "we have misperceived ourselves since Columbus."[41] But it is also true that criollos and affranchis in Latin America developed a nativism and a nationalism ahead of much of Europe. Why? The answer lies partly in the hybridization of cultures and the deracialization of race in which race remains a factor, and in which the concept of miscegenation acquired a positive cast that it did not acquire in North America. (In the United States, nationalism takes the form of white ethnic chauvinism.) Haitian racial definitions were borrowed from those of northern Europe (France, England, and the United States), and *mulâtres* became "blacks" except when at home, where they became *blan peyi* (local whites)—elsewhere, *blanco de la tierra* (in Spanish), and *branco da terra* (in Portuguese)—raising the specter of caste-like social arrangements and quasi-ethnic differences.[42] *Limpieza de sangre* (purity of blood, clean blood), is but an approximation of desire. New peoples arose in the colonial contexts, different from the lower classes—the raw material for new sociocultural formations. The latter retained a "different" outlook, and their original culture evolved as counterculture.

In religion, the other primary cultural element defining nationality, Haiti's early break with Africa not so paradoxically reinforced Africanisms in which content remains strangely familiar today to Ewe and Fon populations. As in Quebec with French Catholicism, that Vodou remained vital to most forms of political resistance cannot be disputed. The armed resistance to the U.S. occupation, which came exclusively from the countryside, was anchored in the national religion, as seen in this exchange in December 1921 between Monseigneur Jean-Marie Jan, the French Roman Catholic Bishop of Cap Haitien, and Senator Medill McCormick, a member of the U.S. Senate commission that investigated the reports of atrocities leveled against the occupation:

Senator: Your comments on the U.S. occupation of Haiti?
Bishop: The occupation was an act of kindness . . .
Senator: The war against the Americans, how do you explain it?
Bishop: The people were pushed to desperation by [American] arbitrariness, injustice and mistreatments. It proclaimed its right to self-defense. Does the United States want to impose Protestantism by force?
Senator: The Washington government will never attack Haiti's [Catholic] faith. Can you provide information on Vodun, its practices and status? Has it diminished since the occupation?
Bishop: It has increased. . . . The greatest cause for this is that the bocors (*bokò*) were the soul of the insurrection.[43]

In the battle between French and U.S. imperialism, the French bishop had expressed the well-founded fear that U.S. missionaries would weaken the French traditional advantage in Haiti. At the same time, Edwin Denby, the U.S. Under Secretary of the Navy, was writing E. O. Watson, the director of the U.S. Council of Churches, that "Haiti is within the sphere of the United States . . . and the Churches and the missionary societies can be of very real help. . . . The Navy Department will be happy to facilitate your work in any way possible."[44] The very serious efforts to convert the Haitian population to American Protestantism should be an indication that others have understood the importance of Vodou. The persecutions orchestrated by many Haitian presidents (although themselves believers in Vodou), following Toussaint L'Ouverture's recognition that resistance and conspiracies could well come from these quarters, should provide similar indications. It seems certain that Haitians have found the millenary religion to be a source of empowerment over the past five centuries. Less a religion than a spiritual discipline sustained by awesome rituals, Vodou, as with Kreyòl, provides a fluidity, a cultural continuum and a bridge between those in the upper and middle classes who follow it for its *magie* (magical powers) and those at the bottom who recognize it as a world view sui generis.

The dichotomies between European and African, slaveowner and slave, brown and black, French and Kreyòl, elite and peasant, and Christian and servant of the Loa were created largely out of the necessity for a small minority to subjugate the majority and maintain control for the primary purpose of acquiring wealth. The gulf created through these juxtapositions would seem to be insurmountable, sapping the very notion of "nation." Having their genesis in colonialism, these bifurcations should be seen not as opposites, but as part of a spectrum and a cultural continuum in which Haitians find themselves. Neither Indian Hinduism, Japanese Shinto, nor Native American spirituality has suffered the indignities born by Vodou. The Japanese context in which Buddhism and Shinto have their spheres is instructive. One acknowledges generally the singular importance that Vodou has on all Haitian arts, although not in other domains such as science and psychology.

Questions regarding the issue of identity have been the concern mainly of elites in Haiti, as have questions regarding the power to define, for as new hybrid cultures were developing from the clash of civilizations in the Americas, no such worries occupied the collective minds of le peuple. The issue becomes important, however, when tied to power and class. It needs to be resolved, furthermore, when liberation—subsuming freedom and democracy—is at stake. One must not in the process jettison the outstanding achievements, norms, and values created in the West that could help in the evolution and development of Haiti. A redefinition of both *form* and *context* must take place, however, based on the ethos of a people neglected far too long. A society in which the haves crush the have-nots can never be democratic nor can it be a community, no matter how numerous the elections. And within the purview of a well-organized state, structure and institutions must be willing to address the well-being of the majority in order to ensure some sense of belonging, a modicum of stability, and social peace. Repression inevitably occurs when oppression remains strong; it also occurs when institutions created by the majority are not respected but are instead targeted for destruction. Democracy in Haiti will be an Africanizing process, as the structures of the state come to reflect the institutions of the nation.[45] One might well assume that ancestral values and norms will not be prized as they resurface in the body politic after a lapse of two centuries. New intellectual paradigms in Haitian scholarship are in the offing.

If Haitian history teaches us anything at all, it is the inordinate love for democracy in a people that has never stopped fighting against those who would deny them that freedom. That history is little understood by scholars outside Haiti; the concept of political culture has hence been misapplied; and democracy, seen as the preserve of Western societies, has been defined improperly.

Notes

1. Boukman Eksperyans, a "roots" (racine) band with a high octane "world (African) beat," is named after him. A bòkò is a Vodou priest in the north of Haiti, or a root doctor or magician, elsewhere.
2. Dantès Bellegarde, *Histoire du Peuple Haïtien* (Port-au-Prince, 1953), 59. Carbonarisme refers to a secret Italian society that met in the woods to plot that country's unification in the nineteenth century under the leadership of Guiseppe Garibaldi (1807–1882).

3. Ibid., 77. It was understandable that enslaved Africans—considered to be property—declared war on property.
4. See Patrick Bellegarde-Smith, *Haiti: The Breached Citadel* (Toronto: Canadian Scholars' Press, 2004).
5. As an aside, when unable to find the solution to an algebraic problem in English, I turned to French. This artifice allowed me to gain a fresh perspective on a given question. Language—and a rich vocabulary—can frame an argument.
6. That Vodou and Kreyòl comprise non-Africanized forms reflects in part an admission of fear and lack of control on the part of a large group subjugated to a powerful smaller group. It also demonstrates the necessity of using "new" words for new things. Additionally, in Vodou, the absolute necessity of hiding what was already a mystical religion demanding initiation came as the result of colonial interdiction and as the recognition of the validity of all spiritual systems. (The relationship between Buddhism and Shinto in Japan or perhaps that between Hinduism and Buddhism serve me well here.) I do, however, accept the validity and reality of a "Creole culture" as defined by anthropologists.
7. There are numerous explanations for the development of Creole languages in general and of Kreyòl specifically. I accept the logic of the arguments made by linguistics scholar Marie-Marcelle Buteau Racine and the research of independent scholar, Thérèse Roumer, on these matters. Why would enslaved Africans born in Africa "forget" the structure of their languages? Kreyòl is not a romance language, as stated in Webster's Dictionary, as Yiddish is Germanic.
8. My experience in West Africa would seem to support this argument, where I stunned Africans by demonstrating "pure" Ewe behavior while I was dancing. Many others have spoken of Kreyòl in similar terms.
9. *Voye pwent* is a feature of Kreyòl that allows a speaker verbally to attack someone indirectly by innuendo. Kreyòl shares these linguistic features with other fairly autonomous African-derived languages and dialects in the Western Hemisphere, notably, black English (or Ebonics) in the United States. "Playin' the dozens," "signifyin'," and "testifyin' " are common examples of kinds of discourse in black English. See the works of Geneva Smitherman.
10. The Kreyòl lexicon is overwhelmingly French although it comprises far-reaching contributions from various African languages, English, American, Spanish, Dutch, and Portuguese. Similarly, although its base is Germanic, English owes much of its vocabulary to Norman French, Latin, and Greek. All Haitians speak Kreyòl; 10 percent know French, and as many or perhaps more, know English. Approximately 5 percent are fluent in Spanish.
11. In the religion these rituals are kept separate, and in various parts of Haiti different rituals dominate. See Robert Farris Thompson, *Flash of the Spirit: African and African-American Art and Philosophy* (New York, 1983).
12. It is common to find Vodou practitioners in most families that purport to despise Vodou. Commonly, one member of an extended family will continue the Vodou rituals for all family members, even among elite families. My personal and academic interests in this subject led me to discover similar behavior among most Haitian presidents and their family members.
13. When a religion is a part of the ethnic patrimony, which is typical in the African setting, it needs no name, since all individuals participate in its rituals at various levels. In Haiti, Vodou (Voodoo, Vaudou) is an "invention" necessitated by the presence of other religions in the landscape.
14. See Frantz Fanon, *The Wretched of the Earth* (New York, 1968); Aimé Césaire, *Discours sur le colonialisme* (Paris, 1955); and Albert Memmi, *Portrait du colonisé, précédé du portrait du colonisateur* (Paris, 1957).

15. This view is factored into the academic discipline of africology, but not necessarily in other, more Eurocentric disciplines. Africology complements other disciplines that are Eurocentric in thrust. The africological subfield in psycho-social inquiry is particularly à propos here.
16. See James M. Blaut, *The Colonizer's Model of the World* (New York, 1993), 97–100.
17. An illustration is found in Euzhan Palcy's film *La Rue des cases nègres* (Sugar Cane Alley), about Martinique in the 1930s, in which the son of a white planter and a mulatto woman with aspirations for her son, turn against the colonial structure altogether, embittered by the "irrationality" of his father's racism. God is the white man, and the God of Christians is represented—as he should be—as a white man.
18. The revolution had been planned by two Roman Catholic priests, Miguel Hidalgo and José-Maria Morelos. They paved the way for priests Camilo Torres (Colombia) and Jean-Bertrand Aristide (Haiti), who later led populist social movements.
19. While U.S. President Ronald Reagan bemoaned the missed opportunity to have resolved the Native American "problem" once and for all, they were not "forcibly" assimilated. They were, in large measure, however, assimilated culturally, racially, and linguistically. They were also the victims of widespread genocide.
20. Official Haitian historiography relates the creation of the blue and red flag in 1803 as symbolizing the union of the blacks (mostly slave, represented by the blue) and affranchis (mostly mulatto, represented by the red), and the motto "L'Union fait la force" ("In unity there is strength"). There are alternative popular explanations of the same event.
21. Louis-Joseph Janvier, *Les Constitutions d'Haiti* (Paris, 1886), 32.
22. Darcy Ribeiro, *The Americas and Civilization* (New York, 1971), 49.
23. See my lengthy treatment of that subject in *In the Shadow of Powers: Dantès Bellegarde in Haitian Social Thought* (Atlantic Highlands, NJ, 1985).
24. Beauvais Lespinasse, *Histoire des affranchis de Saint-Domingue* (Paris, 1882), 15–16.
25. See Brenda Gayle Plummer, *Haiti and the Great Powers* (Baton Rouge, 1988), 41–66.
26. See the analysis in Bellegarde-Smith, *In the Shadow of Powers*, 30–53.
27. Roger Gaillard, *Les Blancs débarquent: la guérilla de Batraville* (Port-au-Prince, 1983), 238. An English text is Hans Schmidt, *The United States Occupation of Haiti* (New Brunswick, NJ, 1971), 135–153.
28. Cited by Germán Arciniegas, *Latin America: A Cultural History* (New York, 1967), 379.
29. Dantès Bellegarde, *Haiti et ses problèmes* (Montreal, 1941), 17.
30. My experience with the major poets Léon Laleau and Jean F. Brierre and the Guyanais Léon-Gontran Damas over the years has shown me how tenuous the African connection was. At the end of their lives, all three recanted their negritude to me, personally.
31. Jean-Paul Sartre had written this in a preface, "Orphée noire," to Léopold Sédar Senghor, *Anthologie de la nouvelle poésie nègre et malgache* (Paris, 1948), xx.
32. The ephemeral presidencies were the following: 1843–1847: Rivière Hérard, Philippe Guérrier, Louis Pierrot, Jean-Baptiste Riché. 1908–1911: Antoine Simon, Cincinatus Leconte, Tancrède Auguste, Michel Oreste, Oreste Zamor, Davilmar Théodore, Vilbrun-Guillaume Sam. 1956–1957: Joseph-Nemours Pierre-Louis, Franck Sylvain, the "collégial," Daniel Fignolé, General Antonio Kébreau. After 1986: General Henri Namphy, Leslie Manigat, "Namphy II," General Prosper Avril, Ertha Pascal-Trouillot—after seven months of Jean-Bertrand Aristide (René Préval)—General Raoul Cédras between 1991 and 1994, with three "distinct" presidents or prime ministers: Emile Jonassaint (and Jean-Jacques Honorat), Marc Bazin, and then Aristide II, with Robert Malval, Smarck Michel, and Claudette Werleigh.
33. See Bellegarde-Smith, *Breached*, 175.
34. Major efforts were undertaken by the French Roman Catholic Church after 1862 and by U.S. Protestant missions from 1922 to the present.

35. The leadership of President François Duvalier in Haiti, President Léopold Sédar Senghor in Senegal, and African-American nationalists buttress my position.
36. At a conference of the Association of Caribbean Studies in Martinique in 1984, Roger Toumson, a Martinican thinker speaking on Caribbean identity, stopped in mid-sentence to exclaim: "Only in the Caribbean could we argue about identity after 500 years of colonization. The issue of identity is alien to the French and to most Europeans."
37. The creation of the Haitian army by the United States in 1916 is a part of that movement. See Sidney W. Mintz, "Preface," in James G. Leyburn, *The Haitian People* (New Haven, 1966), v–xxxvi.
38. This is translated as "government-at-night," referring to secret neo-African societies that "govern" the country when the Westernized urban authorities are asleep. Brazilians have a saying, "the country grows when the government is asleep."
39. My impression is that, when given a chance, Haitians become fluent more easily in Spanish and English than in French, since these languages do not carry the psychological baggage of the last.
40. To speak in sharp, crisp Parisian tones versus speaking flatly, in grammatically correct but Haitian-French intonations.
41. Conversations with Jan Carew, Manaus, Brazil (July 1995).
42. See the analysis in Bellegarde-Smith, "Rum as Cognac: Fluidity of an Ethnocultural Crisis—Haiti," *Kaleidoscope II* (Milwaukee, 1994), 13–18.
43. Jean-Marie Jan, *Collecta III* (Port-au-Prince, 1995), 340–344.
44. Ibid., 350–351.
45. Bellegarde-Smith, *In the Shadow of Powers.*

Chapter 8

The Haiti-New Orleans Vodou Connection: Zora Neale Hurston as Initiate-Observer

Richard Brent Turner

The Negro has not been Christianized as extensively as is generally believed. The great masses are still standing before their pagan altars and calling old gods by new names.

—*Hurston*, Sanctified Church

Introduction

This chapter will analyze Hurston's journey as an initiate-observer of Vodou and her "introspection into the mystery" of the religion as a Vodou adept serving the spirits. This is an extraordinary religious narrative of an initiate and an observer, interpreting the key themes of Vodou, tapping the magical-spiritual wisdom of her elders and ancestors and recording what Haitian adepts call the "*konesans*"—the simple and complex esoteric spiritual knowledge of ordinary Black folk who create transformative healing rituals in African-diasporic communities.

The chapter explores the religious significance of Zora Neale Hurston's initiation and fieldwork in Vodou in the 1920s as the locus for the construction of a Haiti-New Orleans African-diasporic cultural identity with provocative historical, spiritual, and artistic linkages between Black communities in Haiti and New Orleans from the nineteenth century to the present. In Hurston's publications on New Orleans Vodou—*Mules and Men* and "Hoodoo in America," *The Journal of American Folklore* (October–December, 1931), there is important primary source material for analysis of this Haiti-New Orleans Vodou connection and its significance for African-American and African-diasporic religious history.[1]

Recently, Hurston's work has been the subject of critical analysis in African-American theological and religious studies. Katie G. Canon analyzes Hurston's utilization of the folk culture and values of the Black church in her novels as a model for "Black Theological ethics" and "the moral wisdom of Black women."[2] In his book *Honoring the Ancestors*, Donald H. Matthews explores Hurston's research methodology as an early model for "the African-centered approach to the interpretation of African-American" cultural, theological, and literary studies.[3] In *Conjuring Culture*, Theophus H. Smith demonstrates how the Bible served as "a magical formulary" in Hurston's folklore research.[4] Finally, Anthony B. Pinn sketches the history of Vodou in Benin, Haiti, and New Orleans and discusses Zora Neale Hurston's fieldwork as a way of rethinking the canon of Black religion in Black theological studies.[5]

Pinn's *Varieties of African-American Religious Experience*, which reveals the "rich diversity of black religious life in America" by focusing on non-Christian "popular religious practices and sites," informs the African-diasporic orientation of this study. The following two canonical issues that orient his research are central for understanding Hurston's unique contribution to Black religious history in her time: (1) The "narrow agenda and resource base of contemporary" black religious studies, and (2) The "contention that African-American religious experience extends beyond . . . [the] institutional and doctrinal history" of Protestant Christianity to include Islam and African-diasporic ancestral traditions.[6] Indeed these key issues were reflected in the religious meaning of Zora Neale Hurston's research and initiation in Vodou in New Orleans and Haiti. She was 70 years ahead of her time in her exploration of the enduring connections between American religions and African-diasporic traditions, and her analysis of the power and richness of urban folk religions and creolized synthetic-religious identities that stand on their own ground. Hurston's interests reflect the current paradigmatic shift in religious studies in a new wave of cutting-edge scholarship, such as the Vincent Wimbush "Research Project on Society, Culture, and Sacred Texts," Walter F. Pitts, *Old Ship of Zion*, Karen McCarthy Brown, *Mama Lola*, and Robert A. Orsi, *Gods of the City*.[7]

Zora Neale Hurston and New Orleans Vodou: The Haitian Concept of Knowledge, *Konesans*

New Orleans is now and has ever been the hoodoo capital of America. Great names in rites that vie with those of Haiti in deeds that keep alive the powers of Africa.

—*Hurston,* Mules and Men, *183*

In 1928, Zora Neale Hurston received a bachelor's degree in anthropology from Barnard College and traveled to New Orleans in August to begin six months of intensive fieldwork among the city's Vodou adepts. Hurston was a student of the renowned anthropologist, Franz Boas at Columbia University. However, her spiritual experiences in New Orleans Vodou transformed her from a mere participant-observer collector of folklore to an initiate-observer, deeply involved in the key themes and esoteric knowledge and rituals of the religion.

Previous studies of Hurston's New Orleans fieldwork have minimized the depth of "her angle of vision" as an initiate to focus on her work exclusively "as anthropological documents . . . and literary texts." Although this chapter does not contest the validity of prior scholarship, it focuses on another provocative subject of analysis that is rooted in the study of Vodou as an African-diasporic religious tradition with deep roots in Haitian spirituality. Here, *Mules and Men* will be analyzed as a parallel expression of the Haitian "*Konesans*"—esoteric knowledge, both simple and complex that originates in the spiritual experiences of initiates and mystics and is expressed intuitively in the healing and magical rituals of the religion. Hurston's record of esoteric knowledge in *Mules and Men* and "Hoodoo in America" (mentioned earlier) is very important as a primary source of information about New Orleans Vodou in the early twentieth century and its connection to Haitian Vodou.[8]

The Haitian scholar Milo Rigaud sheds light on Vodou's magical-esoteric knowledge, which is embedded in the meaning and the idea of the word:

> Voodoo encompasses an exceedingly complex religion and magic with complicated rituals and symbols that have developed for thousands of years . . . everything essential to the knowledge of the mystery is implicit in this word . . . Vo means "introspection" and Du means "into the unknown." Those who indulge in this "introspection" into the mystere (mystery) will comprehend not only the Voodoo gods, but also the souls of those who are the adepts and the servants of these gods. This is the only way in which the fruitful practice of the rites is possible to produce supernaturally extraordinary phenomena or magic.[9]

Certainly, in *Mules and Men*, Zora Neale Hurston views magic as the centerpiece of New Orleans Vodou's esoteric knowledge and introspection into the mysteries of the cosmos. She writes at the beginning of the Hoodoo section of the book:

> Belief in magic is older than writing. The way we tell it, Hoodoo started way back there before everything. Six days of magic spells and mighty words and the world with its elements above and below was made. And now God is leaning back taking a seventh day rest.[10]

This esoteric ontological statement situates New Orleans Vodou in the realm of the profound and ancient mysteries of the religion that Haitian adepts such as contemporary visual artist, André Pierre, acknowledge as "a world created by magic":

> The Vodou religion is before all other religions. It is more ancient than Christ. It is the first religion of the Earth. It is the creation of the World. The world is created by Vodou. The world is created by magic. The first magician is God who created people with his own hands from the dust of the Earth. People originated by magic in all countries of the world. No one lives of the flesh . . . The spirits of Vodou are the limbs of God.[11]

In Hurston's religious narrative in *Mules and Men*, Moses is the first and most powerful Vodou spirit because he received God's magic rod and acquired knowledge of ten of God's powerful words from the snake that rested under God's feet during the world's creation. This narrative is not just a folkloric transformation of a biblical story into "a conjure story" as Theophus Smith writes, but it is another path through which Hurston establishes the Haiti-New Orleans Vodou connection in the realm of "*konesans*"—esoteric knowledge.[12] In her later work, *Tell My Horse*, she explains the religious meaning of the folk stories of Moses's magic as "Damballah Ouedo . . . the supreme Mystere" whose "signature is the serpent." In New Orleans Vodou, from the time of the nineteenth-century Vodou Queen Marie Laveau to the present, Danbala Wedo has been the constant and sometimes the only *Lwa* involved in rituals as "the ancient sky father [arched] across the sky as a snake beside his rainbow wife, Ayida Wedo. He is the origin of life, and the ancient source of wisdom." According to Hurston, in Haitian Vodou, however, Danbala Wedo always has been one of the pantheon of *Lwa:*

> In the [Haitian] Voodoo temple or peristyle, the place of Damballah, there also must also the places of Legba, Ogun, Loco, the cross of Guede who is the messenger of the gods, of Erzulie, Mademoiselle Brigitte and brave Guede. Damballah resides within the snake on the altar in the midst of all these objects.[13]

The current scholarly consensus established in the research of Patrick Bellegarde-Smith and other groundbreaking studies of Haitian Vodou is that there is a synthetic not a syncretic relationship between Vodou as a powerful African-diasporic religion of resistance to Western hegemony and Christianity.[14] Thus, Vodou stands on its own ground and has not been absorbed by mainstream Christianity. Hurston acknowledged this important synthetic quality of Vodou in New Orleans and Haiti. She writes in *Tell My Horse*,

> The Haitian gods . . . are not the Catholic calendar of saints done over in black . . . This has been said over and over in print because the adepts have been seen buying the lithographs of saints, but this is done because they wish some visual representation of the invisible ones, and as yet no Haitian artist has given them an interpretation or concept of the Lwa. But even the most illiterate peasant knows that the picture of the saint is only an approximation of the Lwa.[15]

Also, in *Mules and Men*, the African ancestral tradition, called hoodoo and conjure by early twentieth-century New Orleans Black folk, is not a mere syncretic footnote in African-American religious history. In Hurston's work the secrecy and the centrality of Vodou in the religious life of New Orleans is part of the esoteric knowledge and power of the religion just as it is in Haiti:

> Hoodoo . . . is burning with all the intensity of a suppressed religion. It has thousands of secret adherents. It adapts itself like Christianity to its local environment, reclaiming some of its borrowed characteristics to itself . . . It is not the accepted theology of the Nation and so believers conceal their faith.[16]

Therefore, in her Vodou trilogy, *Mules and Men*, *The Journal of American Folklore* (1931 issue), and *Tell My Horse*, New Orleans is mapped as a significant religious site, not only because of its relationship to Black southern conjure and root work, but more importantly for its connection to the creolized synthetic fragments of Haitian culture and spirituality that lie deep in its history. Thus New Orleans is written large on the map of Black religion as a magical African-diasporic city, which, like Haiti, has successfully resisted mainstream Christianity's efforts to absorb its esoteric African ancestral knowledge.[17]

Haitian and New Orleans Vodou: Early Historical Connections

What were the important aspects of the early historical encounters between Haitian and New Orleans Vodou in the eighteenth and nineteenth centuries? This subject is the beginning of the esoteric knowledge of New Orleans Vodou that Hurston presents in "Hoodoo in America":

> [Vodou] has had its highest development along the Gulf Coast, particularly in the city of New Orleans and in the surrounding country. It was these regions that were settled by the Haytian emigres at the time of the overthrow of French rule in Haiti by L'Ouverture. Thousands of mulattoes and blacks, along with their white ex-masters were driven out, and the nearest French refuge was the province of Louisiana. They brought with them their hoodoo rituals, modified of course by contact with white civilization and the Catholic Church, but predominately African. These island Negroes had retained far more of their West African background than the continental blacks.[18]

Approximately 12,000 Haitian immigrants arrived in New Orleans from the 1790s until 1810, seeking refuge from the Black revolution in the former French colony, St.-Domingue. These immigrants arrived in a city that already had a distinctive African heritage and religion. According to Gwendolyn Midlo Hall, the Afro-Creole imprint on New Orleans culture was dynamic and constant in the 1700s under French and Spanish rule. Colonial New Orleans culture revolved around the Louisiana Creole language that slaves from Senegambia created in the early 1700s. These Afro-Creole slaves also created rich religious, musical, and folkloric traditions that were influenced by significant cultural and social interactions with Choctaw, Houma, and Chickasaw Indians; French and Canadian Catholic settlers; and pirates. In the seaport city of New Orleans, racial and cultural mixture was widespread and institutionalized but New Orleans still remained the most Africanized slave culture in the United States. Vodou, brought directly to the city from West Africa by Yoruba and Fon slaves from the Kongo and the Bight of Benin, was certainly a major factor in this dynamic African-diasporic culture in the eighteenth century.[19]

Beginning in the late 1750s, New Orleans slaves of Bambara, Mandingo, Wolof, Fulbe, Fon, and Yoruba origin, racially mixed free people of color, and Native Americans gathered every Sunday for communal trading and recreation in the Place des Nègres, an open area behind the city that had earlier been a sacred site for the Houma Indian corn feasts. This site, which was later called Congo Square, was noted for its interaction of African and Indian communities and African dances, drumming, and songs that were performed by hundreds of New Orleanians of African descent until the American Civil War era. Certainly, Congo Square was the most important site in the United States for the public performance of African dances, such as the *bamboula, calinda, coujaille*, and *pile chactas*, which were related to Vodou rites, and it was the marketplace for Afro-Creole Vodou charms that were called *gris-gris* in New Orleans, a term from the Mende language of the Mandingo and Bambara people.[20]

The *Code Noir*, "The Collection of edicts, declarations, and decrees concerning the discipline and the commerce of Negro slaves of the Islands of French America," signed by Louis XIV in 1685, incorporated people of African descent in St.-Dominigue and New Orleans into a common French Catholic colonial culture that allowed the space for the liminal African rituals of Congo Square. Various articles of the *Code Noir* required owners to baptize and bury their slaves according to the Catholic religion and to excuse them from work on Sundays, holy days, and for funerals.

The *Code Noir's* legal permission for African drumming in Congo Square was central to New Orleans' position as the Vodou capital of the United States and its attraction to Haitian immigrants who practiced Vodou.[21] Bobby Joe Neeley believes that in the context of the mass exodus of Vodou adepts from Haiti to New Orleans in the early 1800s, "Voodooism and its sacred dance were institutionalized/acculturated in Congo Square."[22] His research sheds light on the importance of the drumming in Congo Square as the means by which the sacred rhythms of the *Lwa* were invoked in the dances.

Furthermore, in the drawings of the English architect Benjamin Henry Latrobe, based on his observations of the Congo Square rituals in 1815, there is evidence of three different kinds of African drums—the Conga and Ogororo from the Yorubas and the "open-staved drum" from Dahomey. Certainly, these drums were used in the public and private Vodou rituals of New Orleans, which were performed by adepts from Haiti and the Crescent City in the nineteenth century.[23]

Indeed, the arrival of thousands of Haitian immigrants, particularly those who were free persons of color and slaves in the late-eighteenth- and early-nineteenth-century New Orleans, created a critical mass of Black French-speaking people who resisted Americanization after the Louisiana Purchase in 1803 and continued the traditions of Afro-Creole Vodou into a golden age of development that ended in the aftermath of the American Civil War in the late 1860s. In spite of the brutalities of slavery that were present in Louisiana as well as the rest of the American South, Black antebellum New Orleans, like Haiti, was predominantly Catholic and Creole, and had a fluid "three-caste racial system: whites, free persons of color, and slaves" that was constructed in a unique multicultural society.[24] As numerous Black Haitians intermarried with Black New Orleanians at the Ursulines Chapel and St. Louis Cathedral, undoubtedly many provocative spiritual exchanges occurred between Haitian and New Orleans Vodou adepts, and these exchanges created the environment for the ascendancy of the famous nineteenth-century New Orleans Vodou Queen Marie Laveau (1801?–1881).[25]

Vodou Queen Marie Laveau and the Haiti-New Orleans Connection

Marie Laveau is the great name of Negro conjure in America.

—*Hurston, "Hoodoo in America," 326*

The ancestral legacy and legendary spiritual powers of Marie Laveau are a central element in Zora Neale Hurston's twentieth-century religious narratives of Vodou. In this section we will briefly sketch the Vodou Queen's connection to Haiti and Hurston's work. An autonomous Vodou priesthood, similar to the Haitian *manbos* and *houngans*, emerged in early-nineteenth-century New Orleans. Certainly the timing of the emergence of the priesthood was connected to the arrival of thousands of Haitian immigrants in the Crescent City and their connection to Vodou. The New Orleans priestesses and priests, called queens and doctors, included Marie Saloppé, Sanité Dédé, Betsey Toledano, Marie Comtesc, Marie Laveau, Marie Glapion (daughter of Marie Laveau), Doctor Jim Alexander, and Doctor John Montane in the nineteenth century.[26]

Marie Laveau, the most important spiritual leader in this group, was a Catholic free woman of color, born in New Orleans probably in 1801. Her connection to Haiti began with her first husband, Jacques Paris, a free man of color, who was born in St.-Dominigue; the couple married in St. Louis Cathedral in 1819. Although the marriage lasted only a few years, Laveau most likely developed a perspective about the power of Vodou as political resistance from her husband's experiences in his native land. Throughout her adult life, she was known in the Black community as a devout Catholic who attended daily mass at St. Louis Cathedral. The Cathedral was a crossroads for the Haiti-New Orleans Vodou connection in the nineteenth century as the majority of the free women of color in the city were members of the congregation.[27] Laveau's Catholic background is an aspect of the ancestral legacy for Hurston's twentieth-century initiation in Vodou. Hurston notes in her narrative that three of the Vodou doctors with whom she studied were "New Orleans, Catholic," and had an ancestral connection to the powers of Marie Laveau.[28]

Ina Fandrich, in her dissertation, "The Mysterious Voodoo Queen Marie Laveaux," has noted that

> [t]his congregation was probably one of most integrated places in the world at the time and thus became a safe haven for the free women of color who soon formed the majority of the parish membership . . . It is not surprising that most New Orleanian Voodoo worshippers—including Marie Laveaux, and many other prominent priestesses, most of whom were free women of color—were also members of this progressive church. The Voodoo religion, whether in its Haitian or its New Orleanian form . . . found a strange home under Catholic auspices where what appeared on the surface to be Christian symbols and rituals was abundantly filled and redefined with African spirit and meanings.[29]

Marie Laveau reigned as the Vodou Queen of New Orleans from 1850 to 1869. Vodou reached its highest point of development as an organized religion in New Orleans under her extraordinary leadership.[30] According to Fandrich, one of the reasons that Laveau was such a powerful priestess is that she was an example of

> the Haitian concept of priesthood . . . [the head of a] spiritual house; i.e. the worship societies similar to the humfor societies in Haiti, and beyond [her] immediate followers . . . [she] served the larger community as consultant in all aspects of life drawing from [her] divinatory skills and as public healer (because of [her] medicinal, psychological, and spiritual knowledge).[31]

Certainly, Laveau's extensive spiritual work among prisoners condemned to death in the New Orleans parish prison, her nursing of yellow fever epidemic victims, as well as her individual spiritual consultations and distribution of *gris-gris* at her house on St. Ann Street made her an extraordinary healer and spiritual head of a codependent extended religious family network in New Orleans that included plantation slaves, maroons, free blacks, mixed-race people of color, and white women. Perhaps more than any other New Orleanian in her time, she was able to penetrate the boundaries of race, class, color, gender, and religion to establish a profound African spiritual presence in New Orleans that would continue to inspire Vodou adepts in Haiti and Louisiana in the twentieth century.[32]

Indeed, in Zora Neale Hurston's initiation under Samuel Thompson, there is evidence of a continuation of the Haitian model of the "spiritual house" or *humfor* society in early-twentieth-century New Orleans. Her narrative noted the numerous initiates who worshiped in Thompson's House and participated in her initiation:

> Many came into the room and performed ceremonial acts. But none spoke to me. Nor could I speak to them while the veil covered my face. Then Samuel entered and all the others retired.[33]

Finally, it was Marie Laveau's altars, rituals, and sacred dances consecrated to the *Lwa*, particularly Danbala Wedo, at her first house on St. Ann Street and at her second house on Lake Pontchartrain, in ceremonies attended only by initiates that created her strongest spiritual connection to Haitian Vodou and inspired Zora Neale Hurston's initiation in the 1920s. According to Hurston,

> Every St. John Eve she use to rise out of the lake with a huge communion candle burning on top of her head and one in each hand . . . as she rose from the bottom of the lake and walked to the shore upon the water.[34]

The sites of Laveau's spiritual work were mapped as sacred ancestral spaces for twentieth-century New Orleans Vodou. Hurston's initiation occurred in Samuel Thompson's House on Rampart Street (which intersects St. Ann Street and is across the street from Congo Square) and in the swamp along Lake Pontchartrain. She introduces these sites as important landmarks in the narrative of her initiation:

> On Thursday morning at eleven I was at the shuttered door of the ancient house. He let me in cheerfully and led me straight to the altar. There were new candles unlit. He

> signalled me to help. We dressed the candles and lit them and set three upon tumblers filled with honey, three filled with syrup, and three with holy water, and set them in a semi-circle upon the altar. A huge bouquet of flowers was in the center. [And several days later] . . .
>
> Samuel led us on a truck . . . until a certain spot was reached . . . the swamp was dismal and damp, but after some stumbly walking we came to a little glade deep in the wood, near the lake.[35]

Laveau's spiritual and political power as a Vodou priestess constructed from the Haitian concept of religious leadership, created a safe haven for New Orleans' Vodou adepts and neutralized the impact of police and Anglo-Protestant persecution of her followers until the end of her leadership in 1860. Thereafter her daughter, Marie Glapion, assumed leadership of the New Orleans Vodou community, but she was never able to reestablish the spiritual or political power that her mother had created. By the end of the nineteenth century, all of the great Vodou doctors and queens were deceased and New Orleans Vodou went underground as the federal, state, and city government and Black and White American Protestant leaders attempted to destroy African-diasporic folk religion in Louisiana. This hostile political environment influenced the practice of New Orleans Vodou in the era of Zora Neale Hurston's initiation.[36]

The Initiation of Zora Neale Hurston: Fragments of the Haitian Model of Initiation in New Orleans Vodou

> *A man or woman becomes a Hoodoo doctor in three ways: by heredity, by serving an apprenticeship under an established practitioner or by the "call" . . . there is general belief that the power can be transmitted, and for this reason most of the old doctors in New Orleans claim kinship with Marie Laveau.*
>
> —*Hurston, "Hoodoo in America,"* 320

The Sunday African-diasporic dances at Congo Square ended in the 1870s as the Americanization of Franco Africans in New Orleans accelerated. At the same time, the status and fortunes of New Orleans Creoles of color declined rapidly in the late nineteenth century because the white American Protestant racial order replaced the tripartite racial system of antebellum Franco-Catholic New Orleans. Although multicultural Black New Orleanians entered the twentieth century compromised significantly by the new American Jim Crow system, they were one-half of the population of the city and thus found new ways to resist the assaults on their African-diasporic culture and to continue to practice Vodou.[37] The "intimate exchange" between Haitian and New Orleans Vodou continued in the old downtown Creole neighborhoods in which Zora Neale Hurston was initiated. According to Ina Fandrich,

> Since boats and secret messages constantly went back and forth between the island [Haiti] and the metropolis in the Mississippi Delta, a secret traffic of Voodoo paraphernalia and supplies . . . might have been among the underground exchange items.[38]

Zora Neale Hurston arrived in New Orleans in 1928 to begin her apprenticeship as a Vodou initiate:

> I was told that I must begin my novitiate. I must sleep for nine nights with my right stocking on. I must have clean thoughts. I must neither defile body nor spirit. Certain monies were necessary for the ceremony. I paid the sum. I was told to be seated before the altar and offer myself with absolute sincerity to the Great One. But once I was seated I was not to utter a sound. When the spirit was through with me I must leave in silence.[39]

As Hurston experienced her initiation, she observed a religion that had undergone several profound changes since the beginning of the twentieth century. These changes are related contextually to the Haiti-New Orleans Vodou connection. Many New Orleans Vodou initiates were compelled to disguise their religion under the veil of spiritualism—"a technique for communication with the dead," because of the American government's campaign to eradicate Vodou in Haiti and Louisiana.[40] The capstone of this campaign was the U.S. occupation of Haiti and persecution of the nation's Vodou leaders from 1915 to 1934, events that were motivated by American imperialism in the Caribbean and Central America. In this context, American stereotypes of Haitian Vodou expressed in sensational images of bloodthirsty orgies and human sacrifices provided part of the rationale for the criminalization of the Vodou religion in Louisiana.[41] Carolyn Morrow Long has summarized the punitive actions that were levied against some Vodou healers in New Orleans:

> Beginning in 1909, federal mail fraud laws were invoked against practitioners who conducted business by mail; conviction of mail fraud carried a penitentiary sentence. The Louisiana Board of Health, under the 1894 Medical Practice Act, prosecuted persons accused of practicing medicine without a license. A 1929 revision imposed a fifty-to-one hundred dollar fine or a sentence of ten to ninety days in the parish prison. In 1897, 1916, and 1924, the city of New Orleans instituted statutes against fortune-telling and obtaining money under false pretenses.[42]

Faced with these penalties, some Protestant Vodou initiates formed a protective shelter in New Orleans' Spiritual churches. Hurston elaborates on this new synthetic layer of New Orleans Vodou: "Hoodooism is in disrepute, and certain of its practices forbidden by law. A spiritualistic name protects the congregation, and is a useful devise of protective coloration." Spiritual churches, such as the Eternal Life Spiritual Church, were first established by Mother Leafy Anderson in 1918. They included aspects of Vodou, Pentecostalism, Spiritualism, and Catholicism, with elaborate altars to ancestors, saints, and spirit guides, as well as worship services that featured female priests and mothers who prophesized and performed healings during ecstatic trances. At the same time, some Catholic Creoles of color disguised their connection to Vodou through spiritualist séances led by mediums who received detailed messages in French from the spirit world. René Grandjean, who was born in France and lived in Haiti for several years, kept written records of these Creole séances in New Orleans.[43]

Despite these restrictions, Zora Neale Hurston continued to experience the stages of her initiation:

> I sat obedient before the altar, shivering unknowingly. I knew when I was dismissed. I rose and turned from the altar. Then Samuel spoke "The spirit says you must bring three snake skins next time." I passed with bowed head . . . At the end of the nine days I returned with the skins and again sat before the altar.
>
> He prepared the skins and placed them before the Great One. He called him and admonished him to enter into the skins and give them life.[44]

Indeed, all of these profound changes in the culture of Black New Orleans and its connection to Haiti produced a new highly secretive magical emphasis in New Orleans Vodou that focused on spiritual work for clients and resistance to the religion's enemies—it was called hoodoo.[45] It was this magical emphases in Vodou that attracted Hurston. She studied with nine different Vodou doctors and experienced six initiations during her months in New Orleans: "When I found out about Turner, I had already studied under five two-headed doctors and had gone thru an initiation ceremony with each. So I asked Turner to take me as a pupil."[46]

In the narratives of her religious experiences several of the doctors were assigned different names in *Mules and Men* and "Hoodoo in America." This was done to protect them from legal penalties. Thus, Luke Turner in *Mules and Men* was also Samuel Thompson in "Hoodoo in America"; Anatol Pierre, in *Mules and Men*, was also Albert Frechard in "Hoodoo in America"; Father Joe Watson, in *Mules and Men*, was also known as Father Sims, in "Hoodoo in America"; Kitty Brown, in *Mules and Men*, was also Ruth Mason in "Hoodoo in America"; and Dr. Duke, Dr. Samuel Jenkins, Dr. Strong, Doctor Grant, and Doctor Barnes were assigned one name in both of the narratives. These doctors were Vodou priests and priestesses[47] and the most powerful of the group, Luke Turner, was a nephew of Marie Laveau, whose spiritual power came from that nineteenth-century ancestral connection. In Turner's memories of his aunt, described in *Mules and Men*, she was New Orleans' most powerful Manbo in the Haitian model of priesthood, whose favorite *Lwa* was Danbala Wedo, symbolized by the snake:

> Time went around pointing out what God had already made. Moses had seen the Burning Bush. Solomon by magic, knowed all wisdom . . . And Marie Laveau was a woman in New Orleans . . . Alexander the great two-headed doctor felt the power in her and so he told her she must come to study with him. Marie . . . [would] rather dance and make love, but one day a rattlesnake come to her in her bedroom and spoke to her. So she went to Alexander and studies but soon she could teach her teacher and the snake stayed with her always . . . People come from all ends to America to get help from her . . . she hold Hoodoo dance in Congo Square every week . . . and everybody dance like they do in Hayti.[48]

In "Hoodoo in America," Hurston assigned Luke Turner the new name of Samuel Thompson and explicated the source of his ancestral wisdom and its roots in Haiti:

> Samuel Thompson is in his seventies, a Catholic Hoodoo doctor of New Orleans. He has a snake skin, which he says is the skin of the great snake that served her [Marie Laveau]

> altar . . . Samuel Thompson always wrapped his snakeskin about him before attempting any serious work. His mother was a Hoodoo worker and her mother before her. He says that his remote ancestors brought the power with them from "the rock" (Africa) and that his forbears lived in Santo Domingo before they came to the region of New Orleans.[49]

In the nineteenth century, powerful Vodou priestesses such as Marie Laveau were spiritual heads of their own *hounfor* societies or sanctuaries. The new restrictions on Vodou adepts in early-twentieth-century New Orleans forced leaders and their initiates to form secret societies such as the Mardi Gras Indian tribes, like the Creole Wild West and the Yellow Pocahontas that continued aspects of Vodou traditions in their sequin art, Creole dance, and musical rituals and secret initiations in downtown neighborhoods. Their traditions were related thematically to the sequin artists of Haiti who create the flags used in Vodou ceremonies. Also the Mardi Gras Indians' suits included sequined pouches that were inspired by Haitian Vodou healers, sometimes called "Congo pacquets," which were embedded with spiritual power.[50]

In Zora Neale Hurston's mystical initiation in Vodou in Samuel Thompson's House, as well as her spiritual work with the other priests and priestesses, there is evidence of a secret society of magic similar to the Haitian secret societies, namely Bizango. The Haitian secret societies were known for their powerful magic and administration of justice. They were "societies of the night . . . bound by oath" and magical rituals at the crossroads of life and death—in the cemeteries and the forests.[51]

Hurston's account of her initiation brings to light some of the themes of these Haitian secret societies as well as some of traditional aspects of initiation in Haitian Vodou such as separation, liminality, incorporation, fasting, abstinence from sexual relations, spiritual bathing in herbs, psychic experience, trance, apprenticeship purification (*couché*, which means to get initiated), acquisition of knowledge of the *Lwa* and magical rituals, and ritual sacrifice. However, the central theme in the final narrative of Hurston's initiation in New Orleans Vodou that follows is a relationship between religion and magic that is also at the heart of Haitian Vodou in the secret societies according to Rachel Beauvoir-Dominique:

> Vodou emerged with a fundamental vision in which religion and magic, though autonomous, nevertheless constitutes a single body . . . each temple, even the most "religious" in outlook, is under the patronage of one or several . . . divinities destined to work, render service and even amass small fortunes for their possessors. During ceremony, these Lwa are summoned, not "worshipped." In exchange for periodic ritual feeding, the Lwa are expected to protect and soothsay through regular sessions of divination. And they must also carry out therapeutic magic: lucrative treatments and exorcisms will fall under the responsibility of the Vodou divinities of which failings are hardly tolerated . . . Ritual, medicine, and further techniques of manipulation intertwine informed by a common vision of self in which maneuverable human spiritual entities flow in relationship with other forces, all intermingling in the essence of a higher power. The vision emerges from a constantly evolving world and the possibility of adaptation, change, and amelioration.[52]

Beauvoir-Dominique's brilliant analysis of the spiritual and social realm of Vodou's magical practices speaks to the powerful spiritual philosophy of social justice, healing, and resistance to domination that continues to connect the rituals of Vodou initiates in Haiti and New Orleans in the twenty-first century. In Hurston's final religious narratives below, we experience the wholeness of being, the codependence of men, women, plants, animals, and the spirits as Vodou initiates work the cosmic energies of the universe in magical healing rituals that affirm the humanity and central importance of ordinary Black folk in spirit work. Although Hurston is silent about the details of her possession-performances and the identities of the *Lwa* (this is esoteric knowledge for initiates only), the altars are the faces of the *Lwa* and the crossroads of a powerful ancestral spirit world, shared by West and Central Africans, Haitians, and New Orleanians in their rituals.[53]

The excerpts from Zora Neale Hurston's initiation narrative that follow are important for students and practitioners of Vodou in Haiti and New Orleans as the first reliable record of a Vodou initiation in the United States:

> Then I rose from the altar and helped prepare the sacred couch, that is, the moccasin hide was fixed to green cloth and spread over the couch in the altar room. My sacred garments were made, including the crown. At three o'clock, naked as I came into the world, I was stretched face downwards with my navel to the serpent and a pitcher of water at my head that my spirit might not wander in search of it, and began my three day search for the favor of the Great One. Three days I must lie silently, that is, my body would be there. My soul would be standing naked before Spirit to see if he would have me.
>
> I had five psychic experiences during those three days and nights. I shall not detail them here; but I knew that I had been accepted before the sixty-nine hours had passed. Strangely enough, I had no sense of hunger—only exaltation. At eleven o'clock on March 19, St. Joseph's Day, I arose and was led through the running water and again stretched upon my face upon the couch. Samuel approached me with a brother on either side of him. One held a small brush with red paint, the other a brush with yellow paint. With ceremony Samuel painted a lightning symbol down my back from my right shoulder to my left hip. This was to be my sign forever. The Great One shall speak to me in storms. I was now dressed in the new clothes, stockings, underwear, dress and veil. After I was dressed a pair of eyes was painted on my cheeks just below my eyes as a sign that I could see in more ways than one. The sun was painted on my forehand.
>
> At high noon I was seated at the splendid altar. It was dressed in the center with a huge communion candle with my name upon it set in sand, five large iced cakes in different colors, a plate of honeyed St. Joseph's bread, a plate of serpent-shaped breads, spinach and egg cakes fried in olive oil, breaded Chinese okra fried in olive oil, roast veal and wine, two huge yellow bouquets, two red bouquets and two white bouquets and thirty-six yellow tapers and a bottle of holy water.
>
> Samuel seated me and stood behind me with his ceremonial hat upon his head, and the crown of power in his hand. "Spirit! I ask you to take her. Do you hear me, Spirit? Will you take her? Spirit, I want you to take her, she is worthy!" He held the crown poised above my head for a full minute. A profound silence held the room. Then he lifted the veil from my face and let it fall behind my head and crowned me with power. He lit my candle from me. But from then on I might be a candle lighter myself.
>
> All the candles were reverently lit. We all sat down and ate the feast. First, a glass of blessed oil was handed me by Samuel. "Drink this without tasting it." I gulped it down and he took the glass from my hand, took a sip of the little that remained. Then he handed it to the brother at his right who did the same, until it went around the table.

> "Eat first the spinach cakes," Samuel exhorted, and we did. Then the meal began. It was full of joy and laughter, even though we knew that the final ceremony waited only for the good hour of twelve midnight.
>
> We all piled into an old Studebaker sedan . . . Out road number 61 we rattled . . . The sheets of typing paper I had been urged to bring were brought out and nine sheets were blessed and my petition written nine times on each sheet by the light from a shaded lantern. The crate containing the black sheep was opened and the sheep led forward into the center for the circle. He stood there dazed while the chant of strange syllables rose. I asked Samuel the words, but he replied that in good time I would know what to say . . . A knife flashed and the sheep dropped to its knees, then fell prone with its mouth open in a weak cry. My petition was thrust into its throat that he might cry it to the Great One. The broom was seized and dipped in the blood from the slit throat and the ground swept vigorously—back and forth, back and forth—the length of the dying sheep. The sweeping went on as long as the blood gushed. Earth, the other of the Great One and us all, had been appeased. With a sharp stick Samuel traced the outline of the sheep then the digging commenced. The sheep was never touched. The ground was dug from under him so that his body dropped down into the hole. He was covered with the nine sheets of paper bearing the petition and the earth heaped upon him. A white candle was set upon the grave and we straggled back to the road and the Studebaker.[54]

Obviously, Zora Neale Hurston's gifts as a writer are exemplified in this moving narrative of her initiation, as well as her professional ability as a fine ethnographer who described the details of the ritual. However, the most striking quality of this narration is Hurston's transformation into a servant of the spirits, one of the true spiritual folk who has decided to enter the liminal territory of Vodou initiation, which re-creates the African ancestral bridge between black communities in Haiti and New Orleans in the twentieth century. As an initiate of Vodou, Hurston has stripped herself naked to shed her Western identities as a writer and anthropologist to assume the serious responsibility of introspection into the African-diasporic spirit world. The majesty and the awe of the Vodou *Lwa* inspire this narrative of initiation as Zora Neale Hurston has been chosen to tap the wisdom of the ancestors and the healing powers of Danbala Wedo and Shango!

Significance of Hurston's Initiation for the African-Diasporic Spirit World

All of this analysis underlines the religious significance of Hurston's narratives as a lived religious experience. This is the first personal account of an encounter with the religion of Vodou on a deeply personal level in the United States. It is also a record of Zora Neale Hurston's courage and insight as a woman of the spirit world. She was the first African-American to put herself out there in the uncharted waters of Vodou initiation and to emerge from her profound experience as an authentically initiated adept with a written record of her spiritual journey.

The spiritual journey of her initiation rites followed the three classic stages of initiation: "separation," "transition," and "incorporation," which were formulated by

Arnold Van Gennep in *The Rites of Passage*.[55] The new sense of folk community and collective consciousness established among the Vodou initiates in early-twentieth-century New Orleans is a source of continuity between Haitian and New Orleans initiation rites. Here Victor Turner's work is instructive.

Building on the ideas of Van Gennep, Victor Turner developed his theory of "ritual anti-structure." In this theory, he focused on the "liminal period" of initiation in which initiates were considered as "betwixt and between" social statuses in society.[56] Turner regarded this period, during which initiates learned new knowledge about the structure of their community "as a [transitional] process, a becoming, and . . . even a transformation." In the midst of the new social energy that emerged among initiates during the liminal period, Turner proposed that "ritual often acted as a form of protest against the existing social structure and contributed to social change."[57]

Connected to this ritual anti-structure was "communitas," a new sense of community generated among initiates during the liminal period. Turner believed that communitas transformed the structure of society by generating new moods, feelings, and bonds among the initiates, which infused and revived the society with new anti-structural values during and after initiation.[58] The ideas of communitas and liminality are important ways of understanding the significance of Zora Neale Hurston's initiation as a transformative ritual that promoted racial self-definition, gender equality, healing, and resistance to negative stereotypes of Vodou adepts in New Orleans and Haiti.

Finally, Hurston's initiation experience is a model for Beauvoir-Dominique's vision of Vodou "in which religion and magic . . . constitute a single body."[59] In the last stage of the rituals in the swamp, the *Lwa* and mother earth are fed with the sacrifice of a black sheep.[60] This is clearly "therapeutic magic" to insure that Hurston's initiation will promote protection, healing, and spiritual power for all the initiates who performed the rites.[61] Here, the provocative connection between Vodou adepts and the spirit world is highlighted in magical "techniques of manipulation" of the cosmic energies of the "Great One," who created the universe.[62]

Conclusion

In the late 1930s, Zora Neale Hurston resumed her research and introspection into the mysteries of Vodou with two years of fieldwork in Haiti and Jamaica. She wrote *Tell My Horse* in 1938. Although she is celebrated primarily as a Harlem Renaissance literary figure and a folklore collector in Florida, her research and religious experience are also important for understanding the profound and enduring connection between Haitian and New Orleans Vodou.[63]

Today, many Black New Orleanians seriously acknowledge the continuity of the Haiti-New Orleans Vodou connection into a new millennium: well-known New Orleans Vodou priestesses, such as Ava Kay Jones, have traveled to Haiti for initiation as Manbos; "Haitian-style" Vodou services are conducted for initiates; and Haitian Vodou supplies are sold in botanicas in the downtown neighborhoods of the city.[64]

Finally, in 1996, the annual New Orleans Jazz and Heritage Festival celebrated the spiritual and musical connections between Haiti and Louisiana by showcasing the music and culture of Haiti in its first International Pavilion. Festival organizers explained their version of the Haiti-New Orleans connection:

> Several Haitian musicians, bands, and dancers are performing at this year's Festivals, expressing the myriad of rhythms that distinguish their homeland: Compas, the dance music genre . . . ; Rara, Carnival music highlighted by colorful parades that are similar to New Orleans' Second Lines; and Racine, a musical genre that enhances Voodoo's rhythms with electric instruments.
>
> Serving as master of ceremonies for the pavilion is Aboudja Derenoncourt, a respected priest and master drummer who leads audiences through an exploration of Haitian history and cultural traditions—traditions that New Orleanians will find complement those of their own.[65]

Notes

1. Zora Neale Hurston, *Mules and Men* (New York: Harper Perennial, 1990); and "Hoodoo in America," *The Journal of American Folklore*, 44, no. 174 (October–December, 1931): 317–418.
2. Katie C. Canon, *Black Womanist Ethics* (Atlanta, GA: Scholars Press, 1988), 15 and back cover.
3. Donald H. Matthews, *Honoring the Ancestors: An African Cultural Interpretation of Black Religion and Literature* (New York: Oxford University Press, 1998), front flap.
4. Theophus H. Smith, *Conjuring Culture: Biblical Formations of Black America* (New York: Oxford University Press, 1994), 6.
5. Anthony B. Pinn, *Varieties of African-American Religious Experience* (Minneapolis, MN: Fortress Press, 1998), 1.
6. Ibid., 1 and back cover.
7. Vincent Wimbush, ed., *African-Americans and the Bible: Sacred Texts and Social Texture* (New York: Continuum International Publishing Group, 2000); Walter F. Pitts, *Old Ship of Zion: The Afro-Baptist Ritual in the African Diaspora*. (New York: Oxford University Press, 1993); Karen McCarthy Brown, *Mama Lola: A Vodou Priestess in Brooklyn* (Berkeley: University of California Press, 1991); Robert A. Orsi, *Gods of the City* (Bloomington, IN: Indiana University Press, 1999).
8. David C. Estes, "The Neo-African Vatican: Zora Neale Hurston's New Orleans," in *Literary New Orleans in the Modern World*, ed. Richard Kennedy (Baton Rouge, LA: Louisiana State University Press, 1998), 67–68; Brown, *Mama Lola*, 384.
9. Milo Rigaud, *Secrets of Voodoo* (San Francisco: City Lights Books, 1985), 8.
10. Hurston, *Mules and Men*, 183.
11. André Pierre, "A World Created by Magic: Extracts from a Conversation with André Pierre," in *Sacred Arts of Haitian Vodou*, ed. Donald J. Consentino (Los Angeles: UCLA Fowler Museum of Cultural History, 1995), xxii.
12. Smith, *Conjuring Culture*, 33.
13. Zora Neale Hurston, *Tell My Horse: Voodoo and Life in Haiti and Jamaica* (New York: Harper and Row Publishers, 1990), 120.

14. Patrick Bellegarde-Smith, ed., *Fragments of Bone: Neo-African Religions in a New World* (Champaign, IL: University of Illinois Press, 2005); Gerdès Fleurant, *Dancing Spirits: Rhythms and Rituals of Haitian Vodun, the Rada Rite* (Westport, CT: Greenwood Press, 1996); Patrick Bellegarde-Smith and Claudine Michel, eds., *Spirit, Myth and Reality in Haitian Vodou* (Indiana: Indiana University Press, 2006).
15. Hurston, *Tell My Horse*, 114.
16. Hurston, *Mules and Men*, 185.
17. Estes, "Neo-African Vatican," 82.
18. Hurston, "Hoodoo in America," 318.
19. Paul F. Lachance, "The Foreign French," in *Creole New Orleans: Race and Americanization*, ed. Arnold R. Hirsch and Joseph Logsdon (Baton Rouge, LA: Louisiana State University Press, 1992), 103–105; Gwendolyn Midlo Hall, "The Formation of Afro-Creole Culture," in *Creole New Orleans: Race and Americanization*, 59–65.
20. Jerah Johnson, *Congo Square in New Orleans* (New Orleans, LA: Louisiana Landmarks Society, 1995), 9–11; Carolyn Morrow Long, *Spiritual Merchants: Religion, Magic, and Commerce* (Knoxville, TN: University of Tennessee Press, 2001), 21; George Washington Cable, "The Dance in Place Congo," *Century Magazine*, vol. 31 (February 1986): 517–532; Michel S. Laguerre, "Voodoo as Religious and Revolutionary Ideology," in *Freeing the Spirit* 4 (1971), 24.
21. Joseph Roach, *Cities of the Dead: Circum-Atlantic Performance* (New York: Columbia University Press, 1996), 56–58, 63–64.
22. Bobby Joe Neeley, "Contemporary Afro-American Voodooism (Black Religion): The Retention and Adaptation of the Ancient African-Egyptian Mystery System," PhD dissertation, University of California, Berkeley, 1988, 449.
23. Johnson, *Congo Square in New Orleans*, 25–26, 31.
24. Lachance, "Foreign French," 101.
25. Ibid., 108–109.
26. Ina Johanna Fandrich, "The Mysterious Voodoo Queen Marie Laveaux: A Study of Female Leadership in Nineteenth-Century New Orleans," PhD dissertation, Temple University, 1994, 82; Long, *Spiritual Merchants*, 44. The Vodou Queen's name is spelled both "Laveaux" and "Laveau" in historical records and scholarship. The author utilizes "Laveau," the spelling that is popular in contemporary Black New Orleans communities.
27. Long, *Spiritual Merchants*, 45–46. After Paris's disappearance Marie Laveau had a long-term relationship and several children with Louis Glapion.
28. Hurston, "Hoodoo in America," 317.
29. Fandrich, "Mysterious Voodoo Queen Marie Laveaux," 166–167.
30. Ibid.
31. Ibid, 82.
32. Daily Picayune, June 16, 1881.
33. Hurston, "Hoodoo in America," 359.
34. Ibid, 327.
35. Ibid, 358, 360.
36. Long, *Spiritual Merchants*, 52.
37. Arnold R. Hirsch and Joseph Logsdon, "Franco-Africans and African-Americans," in Hirsch and Logsdon, *Creole New Orleans: Race and Americanization*, 190.
38. Fandrich, "Mysterious Voodoo Queen Marie Laveaux," 257.
39. Hurston, "Hoodoo in America," 358.
40. Long, *Spiritual Merchants*, 53. Ibid., 319.
41. Laennec Hurbon, "American Fantasy and Haitian Vodou," in Consentino, *Sacred Arts of Haitian Vodou*, 182–187.

42. Long, *Spiritual Merchants*, 53.
43. Ibid, 54. Hurston, "Hoodoo in America," 319. Claude F. Jacobs and Andrew J. Kaslow, *The Spiritual Churches of New Orleans: Origins, Beliefs, and Rituals of an African-American Religion* (Knoxville, TN: University of Tennessee Press, 2001); Jason Berry, *The Spirit of Black Hawk: A Mystery of Africans and Indians* (Jackson, MS: University Press of Mississippi, 1995); Marcus Bruce Christian, "Manuscript for A Black History of Louisiana," in *The Marcus Bruce Christian Collection* (Earl K. Long Library: University of New Orleans), and *Rene Grandjean Collection* (Earl K. Long Library: University of New Orleans).
44. Hurston, "Hoodoo in America," 358–359.
45. Long, *Spiritual Merchants*, 55.
46. Hurston, *Mules and Men*, 191.
47. Hurston, "Hoodoo in America," 317.
48. Hurston, *Mules and Men*, 192–193.
49. Hurston, "Hoodoo in America," 357.
50. Tina Girouard, *Sequin Artists of Haiti* (Port-au-Prince, Haiti: Haiti Arts, 1994); Robert Farris Thompson, *Face of the Gods: Art and Altars of Africa and the African-Americas* (New York: Museum for African Art, 1993), 28.
51. Donald J. Consentino, "Bizango Altar," in Consentino, *Sacred Arts of Haitian Vodou*, 302.
52. Rachel Beauvoir-Dominique, "Underground Realms of Being: Vodou Magic," in Consentino, *Sacred Arts of Haitian Vodou*, 166.
53. Thompson, *Face of the Gods*, 21.
54. Hurston, "Hoodoo in America," 358–360. There is also provocative analysis of Zora Neale Hurston's initiation and spiritual work in Robert E. Hemenway and Zora Neale Hurston, *A Literary Biography* (Chicago, IL: University of Illinois Press, 1977); Alice Walker, ed., *I Love Myself When I am Laughing . . . And Then Again When I Am Looking Mean and Impressive: A Zora Neale Hurston Reader.* (New York: Feminist Press at the City University of New York, 1979); Houston A. Baker, Jr., *Workings of the Spirit: The Poetic of Afro-American Women's Writing* (Chicago: University of Chicago Press, 1991); Cheryl A. Wall, "Mules and Men and Women: Zora Neale Hurston's Strategies of Narration and Visions of Female Empowerment," in *Critical Essays on Zora Neale Hurston*, ed. Gloria L. Cronin (New York: C. K. Hall, 1998).
55. Arnold Van Gennep, *The Rites of Passage* (Chicago: University of Chicago Press, 1966), 11.
56. Victor Turner, *The Forest of Symbols: Aspects of Ndembu Ritual* (Ithaca, NY: Cornell University Press, 1957), 93, 94.
57. Bobby C. Alexander, *Victor Turner Revisited: Ritual as Social Change* (Atlanta: Scholars Press, 1991), 1.
58. Victor Turner, *The Ritual Process: Structure and Anti-Structure* (Chicago: Aldine Publishing, 1969), viii.
59. Beauvoir-Dominique. "Underground Realms of Being," 166.
60. Hurston, "Hoodoo in America," 360.
61. Beauvoir-Dominique, "Underground Realms of Being," 166.
62. Hurston, "Hoodoo in America," 360.
63. Hurston, *Tell My Horse*.
64. Ava Kay Jones, interview with author, December 1997, New Orleans, LA. Long, *Spiritual Merchants*, 69. F & F Candle Shop on 801 North Broad Avenue is the largest botanica in New Orleans.
65. Linda Bays Powers, "Haiti: Waves of Sound and Rhythm," in *The 1996 Official New Orleans Jazz and Heritage Festival Program Guide Book*, 36.

Chapter 9

Water in Their Eyes, Dust on Their Land: Heat and Illness in a Haitian Town

Pierre Minn

Author: "Does heat make people sick?"
Informant: "Yes, there's no water."

The man who spoke these words gave his answer with a slight shrug, as if it were the most obvious response to my question. As I began to understand his statement, I was struck by the enormity of what had previously seemed just another discomfort of living in a poor, tropical country. This chapter is about the concept of *chalè*, or "heat" as it appears in the medical discourses of rural Haitians. The term *chalè* is used to refer to the physical phenomenon of heat as well as to an illness that affects various parts of the body and is thought to be caused primarily by overexposure to heat and physical labor. Drawing from interviews conducted with 30 residents of Bèlans, a coastal town in northern Haiti, I will argue that an understanding of the illness *chalè* requires recognition of environmental factors in the analysis of illness narratives.[1] Environmental factors such as contact with biomedically recognized pathogens and toxic pollutants or the spread of previously isolated micro-organisms have already been recognized as having an impact on human health, but environmental factors must also be considered when examining the lived experiences of illness. Rural Haitians generally do not attribute *chalè* to biomedical causes or social dynamics, but focus instead on environmental problems when describing the illness. Although from an outsider's perspective, one can point to the social roots of poverty, underdevelopment, and poor health, one must not overlook how Haitians link *chalè* to factors as immediate and obvious as a dry riverbed and the sun's blazing heat.

This study is intended primarily as a descriptive chapter. After a brief introduction to the field site, methods of data collection, and humoral classification, I will describe the causes, symptoms, and treatments of the illness. A short section devoted to the biomedical perspective on *chalè* includes opinions expressed by two nurses who work in the town's clinic. I will then describe how the residents of Bèlans connect *chalè* to the broader environmental problems that affect the majority of rural Haitians—namely, drought and land degradation, arduous labor conditions, and extreme poverty. Finally, I will argue that three categories commonly used when studying illnesses cross-culturally ("folk illness," "culture-bound syndrome," and "metaphor") are inappropriate for *chalè* because they emphasize the social dynamics surrounding illness and dismiss physiological, objectively measurable factors. To understand this illness in its appropriate context, it must be examined in a way that acknowledges the importance of the physical environment in which it is experienced.

In recent years, social scientists have shown that health includes more than the biological processes of individual bodies. Medical anthropologists have established distinctions between illness and disease that have influenced how anthropologists theorize about health and medicine. Kleinman writes, "By invoking the term illness, I mean to conjure up the innately human experience of symptoms and suffering. Illness refers to how the sick person and members of the family or wider social network perceive, live with, and respond to symptoms and disability" (Kleinman 1988, 2–3). In contrast, disease is "the problem from the practitioner's perspective. In the narrow biological terms of the biomedical model, this means that disease is reconfigured only as an alteration in biological structure or functioning" (1988, 5–6). I have chosen to refer to *chalè* using the term "illness" because of its emphasis on the lived experience of health. The term is limited, however, in its capacity to convey how *chalè* affects and interacts with the physical environment beyond the bodies that experience it.

Background

Bèlans and its Inhabitants

Bèlans is situated about 25 miles west of Okap (Cap-Haïtien) on Haiti's north coast. The town is on a peninsula and lies alongside a small river that joins the ocean at the edge of the town. Bèlans itself is relatively flat, but the terrain surrounding it is extremely rocky and mountainous. The town has no paved roads, and its streets become filled with muddy pits after rain. A few households own generators, but the vast majority of residents have no access to electricity. A system of pipes collects runoff from the hills and supplies water to several homes and two public spigots, which provide water for drinking and cooking. The pipes often dry up because of drought or damage, and people are forced to fetch water from wells or springs. The majority of the residents bathe in the river, which has served the water needs of thousands of people living in the mountains south of the town by the time it reaches Bèlans. Aside from a small number of pit latrines, there is no sanitation system, and the town is crossed by small canals filled with raw sewage. People also defecate on the banks of the river, which is dark with filth by the time it reaches the ocean.

According to a 1995 government report, Bèlans has a population of 6,500, and its surrounding "communal section" is home to some 60,000 people (Republique d'Haïti, Departement du Nord 1995).[2] Most of the town's residents are dependent on farming as their main source of income. Residents own small plots of lands in the hills surrounding the town, where they grow plantains, bananas, breadfruit, mangoes, and other crops. A few farmers cultivate rice, peanuts, and sugarcane in more level areas, but level land is scarce. Others in the town earn their living as fishermen, metalworkers, masons, vendors, teachers, and furniture makers. There is an open market in the center of Bèlans where people sell staple foods every day from early morning to early evening.

Twice a week, large market days attract traveling vendors from the surrounding areas who come to sell kitchen utensils, used clothing, and other dry goods. A few families also run small stores stocked with pencils, notebooks, insect repellent, batteries, and specialty food items such as crackers, candy, and cheese. Those with refrigerators sell cold beverages and margarine.

Health conditions in Bèlans are marked by the area's extreme poverty. Malnutrition is widespread, as are diseases such as malaria, typhus, and parasitic infections. The use of medicinal plants is common and usually serves as the primary form of care for a wide variety of ailments. There is a government-run hospital just outside of town that employs two Haitian doctors. The building is in a state of neglect, however, and lacks basic supplies such as gauze, antiseptics, and hospital beds. There is also a small clinic in the center of town, which is funded by an American development organization and run by a Haitian staff of two nurses, a lab technician, and a pharmacist.

Research Methodology

I first traveled to Bèlans in the summer of 1997 under the auspices of the development organization that built and currently funds the town's clinic. The organization's goals are to work in solidarity with the people of Bèlans to improve conditions for those who live in the town and its surrounding areas.

During my first seven-week stay in Bèlans, I worked with the group's education outreach program and organized classes for students and teachers in various disciplines. During this time, I became interested in the concept of *chalè*, after observing the use by local residents of a great number and variety of medicinal plants in the area surrounding the town. In explaining a plant's usefulness, informants often described it as being "good for *fyèv*" (fever) or "good for *chalè*." Although many plants could be used to treat both *fyèv* and *chalè*, the illnesses were always listed as two separate conditions, and some plants were good for only one of the two. Whereas *fyèv* was described in terms of chills and generalized aches and pains, *chalè* was more localized, often in the stomach.

I returned to Bèlans in the summer of 1998 to gain a better understanding of these categorizations. During the second stay, which lasted five weeks, I conducted 30 interviews, which consisted of open-ended questions, and usually took place in informants' homes. Few interviews were prescheduled, but because I had already established a visible presence in the town, no one seemed surprised by my questions. No request for an interview was refused; rather, I was impressed by the informants' willingness to share information and details. The 15 female and 15 male informants ranged in age from nine to late seventies. In choosing informants, I tried to choose individuals from a wide variety of occupational backgrounds, who would have disparate experiences and lifestyles. The group included farmers, students, homemakers, carpenters, merchants, tailors, fishermen, teachers, and others. All were residents of Bèlans, and almost all had always lived in the town or its surrounding areas. A few had spent brief periods abroad, working either in the Dominican Republic cutting sugarcane or in the Bahamas as domestic servants and construction workers, but most had never left Haiti. Although I make no attempt to present my group of respondents as a random sample, their disparate demographic attributes, especially in terms of age and occupation, combined with the overall uniformity of their responses convinces me that the information they gave me is representative of Bèlans' general population. I also interviewed the two nurses who staff the town's clinic to evaluate the local biomedical response to *chalè*. (All interviews were conducted in Haitian Creole, without a translator.)

The Illness Chalè

Origins

The Haitian Creole term *chalè* derives from the French word *chaleur* and can be translated literally as "heat." The word is used as a noun, and its corresponding adjective is *cho* (hot). The term *chalè* poses complications because it is used to refer both to heat as an elemental force and to the illness caused by heat. For clarity, I will use the English word "heat" to refer to the elemental force, and "*chalè*" when speaking of the illness. It is questionable to what extent this distinction between the two concepts exists for the people of Bèlans, since many believe that heat simply enters the body and makes one sick. Some people used the term *maladi chalè*, or "heat illness," which indicates that some separation of the concepts exists.[3]

At its most basic level, *chalè* can be described as an illness rooted in Greek-derived humoral pathology. The traditional Greek categorizations of hot, dry, cold, and damp have been preserved in Latin America with the damp and dry categories losing importance to create a binary opposition between the concepts of hot and cold. "The basic outline of Latin American humoral medicine as described in . . . ethnographic accounts is simple . . . all foods, all herbal and other remedies, and many other substances as well (such as iron), have a metaphoric 'quality'. . . a humoral value of 'hot' or 'cold' (and occasionally 'temperate') that serves primarily to distinguish classes of foods and remedies" (Foster 1994, 2). Although humoral classifications are most prevalent in Spanish-speaking Latin America, they are also present in the French Caribbean and Haiti. Foster argues, "[In] Haiti, during its century of French domination . . . it is reasonable to assume that French humoral ideas also played a role in implanting the Hot-Cold concepts revealed by contemporary anthropological research" (176).[4] Several studies have confirmed the existence of humoral classification among Haitians and Haitian-Americans (DeSantis and Thomas 1990; Wiese 1976).[5]

Chalè's Symptoms and Treatments

In Bèlans, when someone says, "*M'gen chalè*," it is assumed that the person has been exposed to an excessive amount of heat, which is causing negative effects within the body. *Chalè* is one of the most common illnesses in Bèlans, and all but one of my informants acknowledged that heat causes illness. (The exception was a nurse in the town's health clinic.) *Chalè* is most often experienced in the stomach and bowels; in the head and eyes; and on the skin. Some informants linked the illness' scope to two or three body parts (usually those mentioned above), but many stated that *chalè* could affect all parts of the body. One woman showed me her foot, the swollen and painful condition of which she attributed to *chalè*. Each area of the body is affected in a particular manner, leading me to conclude that the illness is not conceived of as one single type of disorder (such as too much blood or torn tissue) but as a painful and uncomfortable deviance from normal functioning.

As stated before, I first noticed *chalè* as a condition about which people spoke in terms of their stomachs and bowels. *Chalè* causes pains in the stomach, which the informants described as sharp and severe. In addition, *chalè* in the stomach produces a series of digestive problems, such as *pouse* (pushing): the inability to have a bowel movement despite great effort.[6] *Chalè* also produces *glè*, a condition in which one's stools become clear and runny, but different from diarrhea. One man described *glè* as "when you kill animals, when you're cleaning their tripes, there's a kind of garbage you can see. That's what it's like." *Glè* was described by one informant as being "like bile." Pains (*pouse*) and *glè* were the generally agreed-upon effects of *chalè* in the stomach, whereas only a few informants mentioned diarrhea and constipation as symptoms.

Treatment of *chalè* in the stomach and bowels consists mostly of *rafrèchi*, or "cooling," which refers to the consumption of cold foods and beverages. It was in this realm that traditional humoral classifications became most evident, because informants listed food with cold properties that are used to cool down *chalè* inside the body. These foods included coconut milk, cassava bread (made from ground manioc), okra, cucumbers, sugar, bananas, and melon. All the foods mentioned by my informants were classified as cool or cold by Jean Wiese's informants in a 1976 study of traditional food behavior in Haiti (Wiese 1976, 196). Cooling beverages are made from the leaves of a variety of plants, such as hibiscus, orange, and soursop (a large tree that produces sweet, pulpy fruits), which are boiled and left to cool.[7] All of these plants and foods are available in Bèlans, although they are outside the staple diet of rice and beans and may be a financial burden for the patient. Medicinal leaves are plentiful and gathered from hillsides by the patients or their family members.

Chalè in the head produces dizziness, headaches, and a sensation that is hot but not necessarily feverish. One man told me, "*Chalè* first works in your head. It makes your head heavy, it gives you something like vertigo. The person gets dizzy." The person may have difficulty seeing because *chalè* can affect the eyes, producing soreness and redness. "It makes you feel as though your eyes are burning, as if they are going to fall out of your head and land on the floor," one informant explained. *Chalè* in the head is treated by *koule tèt*, whereby the patient douses the head with cool water into which medicinal leaves may have been infused. It is also common to tie leaves onto one's head with a damp cloth, so that their cooling properties may be absorbed through the skin. One woman told me that after eating a banana, "you can tie it [the peel] to your head with a rag, and it will pull out all the heat. You'll see, after about fifteen to twenty minutes, the peel is almost dry. It has pulled out so much heat, the heat was so strong that the peel has a totally different appearance."

The final major manifestation of *chalè* is on the skin. This condition is the most obvious form of the illness and one of the most uncomfortable. *Chalè* on the skin usually appears in the forms of *bouton* (bumps), which can occur on any part of the body. They are generally small in size and may be filled with pus or clear liquid. It can also appear as a rash over a larger area of skin. *Chalè* on the skin can last anywhere from a single day to long periods of time and is extremely itchy and painful. A popular name for the condition is *gran difè chalè* (big fiery heat). This term is also used to refer to *chalè* in the genitals, described below. *Chalè* on the skin occurs frequently among infants. One young woman, showing me her baby son, said, "You can see that he's got a bunch of little bumps all over his body, that have water in them. Some are big; sometimes they get bigger. They are usually pink. The heat does that. Sometimes they have a lot of water in them. When you pop them, they leave a mark." Bathing is the most common treatment for *chalè* on the skin. In addition, informants described a remedy that consists of mixing starch (usually made from manioc) with *klèren* (a strong form of rum) and rubbing it on the afflicted parts.

Aside from these major symptoms, individual informants spoke of many other manifestations of *chalè*. A few mentioned *chalè* in the genitals, which can affect both males and females, causing the urethra to burn and one's urine to become darker. This condition was referred to by the same name as the skin rash, *gran difè chalè*, and is generally believed to be caused by sitting on something very hot, such as a metal chair or concrete stoop warmed by the sun. Some informants also mentioned that *chalè* in the genitals was sexually transmitted. This form of *chalè* was identified as very serious and deadly if not treated by biomedical remedies. Other effects of *chalè* include hair loss, hysteria, swelling in the stomach, and difficulty in breathing. In all cases, *chalè* was described as an invasive force that originates outside the body and affects the person negatively.

Causes, Consequences, and Classifications

Chalè is primarily caused by overexposure to heat, which is believed to create disturbances within the body. The sun was identified as the primary source of heat but others included hot rooms such as kitchens or crowded bedrooms and small charcoal braziers used for cooking. A person becomes sick by spending too much time exposed to one or more of these sources. Being in the sun for an extended period of time, sitting close to a stove while cooking, and ironing clothes were all frequently mentioned as common ways of being exposed to heat and getting *chalè*. Individual informants told me of other ways one could get the illness: by driving a car, for example, because of proximity to a hot motor, or by welding metal. Surprisingly, not a single informant connected *chalè* in the stomach with ingesting hot foods or beverages.[8]

The other principal cause of *chalè* is physical activity, which not only heats up the body but also weakens its resistance to heat. Activities such as walking or climbing a hillside, carrying heavy loads, construction work, and playing soccer were all described as activities that caused people to become ill. Sometimes this type of activity is thought to preempt the presence of a heat source which causes illness. When I asked a man if it were possible to get *chalè* even when the weather wasn't warm, he responded, "It depends on the activity of that person. There are some people who live here, in Bèlans, but who go to school in Sen Jak [a nearby town]. That means that every day they must go there and every afternoon come back. If they do that every day during the year, they will feel something. Even if it's not hot, it (*chalè*) will show itself. *Chalè* will manifest itself in that person."

Generally, however, heat and *chalè* are thought of in terms of seasons. "Heat is something that has its time. It's something that comes by season. There are times when it's hot and times when it's not hot. During the hot months, the heat rises, and when it goes down, the cold comes in." The *epok chalè*, or "heat season," lasts from May to September with the winter months tending to be much cooler. Informants were divided when asked if it were possible for people to have *chalè* during the winter months. Most agreed that it was more likely to occur during the summer. "In the hot months, that's when it hits the body most. When it's cold, it's not too much, it's more in the hot months," one woman told me. When I asked another informant, however, if it were possible to get *chalè* in January or February, he replied, "It's possible. As soon as you tire out your body, you can get it. You can get it anytime. Whenever a person is tiring their body, they can get it." As stated above, physical labor can make the presence of heat superfluous in making a person ill.

Aside from these two principal causes of *chalè*, another theme that emerged from the interviews was the question of hygiene. I was told, "If a person is in filth, if they are in dirt, they don't want to wash, they don't want to bathe, that'll cause it also." Another man explained it this way, "If you don't take good care of your body, you have little holes in your skin, called pores, there's a lot of dirt there, if you don't take care of your body, *chalè* will kill you quicker." This man was the only informant to mention pores. Another statement concerning hygiene was made by two young boys, who were the only informants to use the word *mikwob* (microbe or germ).

Author: Do you get *chalè* sometimes?
Informant: Yes.
Author: Where?
Informant 1: In our house. We take off our shirts at night. We like to wash so that we don't get germs.
Informant 2: So that we don't have dirt stuck on us.

All of the informants who mentioned bathing and hygiene spoke of them as factors that could be controlled, and implied that failure to maintain cleanliness was caused by a lack of desire or effort. However, plentiful dust and lack of water make it difficult for anyone to maintain strict standards of hygiene. Many people spoke of excessive sweating during the summer months as well as the dust created by high temperatures and scarce rainfall. One man told me that someone herding livestock might get the illness if the animals passed gas and the person inhaled the fumes. There is an association between *chalè* and discomfort: the sweltering heat and ever-present sweat and dirt combine to create an environment that makes people ill.

Chalè is a relatively commonplace condition, and, ordinarily, people rarely consider it worthy of a trip to the clinic or hospital. In general, people in Bèlans treat most of their ailments with home remedies and medicinal leaves, turning to biomedical treatments only when their first attempts fail. Most people asserted that biomedical treatments were available for *chalè*, but few were able to name specific medicines. Those mentioned were antibiotics (as a blanket term), ampicillin, skin creams (one woman mentioned "Calandrel," presumably a calamine-based lotion), and soaps. The hospital and clinic both charge fees for visits and medications, and although both institutions are heavily subsidized, the home remedies for *chalè* remain more economically advantageous. In addition, *chalè* on its own is simply not considered a dangerous illness. Informants emphasized that it was not a *gwo maladi* (big illness), and few believed that it is lethal. *Chalè* is thought to exacerbate other sicknesses, though. As one informant put it, "*Chalè* can't kill people. If a person is already sick, *chalè* can kill. If the person already has another sickness, and they get *chalè*, they might die, but it's not *chalè* that killed them." There is also the threat that *chalè* will make the patient more susceptible to other, more serious illnesses or that *chalè* itself can evolve into a more dangerous ailment, such as a fever or infection.

Scholars have recognized the tendency of Haitians to incorporate biomedical concepts into preexisting models of illness and health. In his book *AIDS and Accusation*, Paul Farmer writes about the ways in which the residents of a small Haitian town came to relate AIDS to a disease with which they had long been familiar—tuberculosis (Farmer, 1994, 117). Another anthropologist, Michel Laguerre, writing on health issues among Haitian Americans, says that "[s]pecific biomedical terms are often used by poorly educated migrants as well, but often with meanings different from those assigned by biomedical professionals" (Laguerre 1981, 201). Similarly, I found during my interviews that biomedical terms and concepts existed with and influenced the concept of *chalè*. The presence of the health clinic and the hospital attest to the fact that despite their relative isolation, the people of Bèlans are in contact with biomedical disease models and remedies. Biomedical terms were most frequently mentioned when speaking of the dangers of untreated *chalè* or what could happen to a patient in a worst-case scenario. As mentioned above, many believed that *chalè* could render one more susceptible to more serious conditions. One man told me that "[*C*]*halè* makes infection unleash itself faster on you".[9] Other informants told me that *chalè* led to fever and frequently mentioned malaria and typhoid. Another condition often related to *chalè* was *tansyon*, or high blood pressure. Several informants, including a biomedically trained nurse in the clinic, informed me that blood pressure rose during hot weather.[10] Because most people in Bèlans resort to the clinic or hospital only if facing a serious condition, it is logical that the more severe cases of *chalè* would be conceived of in biomedical terms.

The Biomedical Perspective

Biomedical Categorization of Chalè's Symptoms

The symptoms of *chalè* can be linked to a wide variety of biomedically recognized diseases and ailments, many of which are related to the conditions of water supplies in Bèlans. Theorists have created a classification system that establishes five categories of infectious diseases linked to water: water-borne infections such as typhus and cholera, whose organisms spread in contaminated water; water-washed infections such as scabies and trachoma that are caused by lack of water for personal hygiene; water-based infections such as schistosomiasis and other worm infections; diseases with water-related insect vectors such as malaria; and infections related to defective sanitation such as hookworm, which move directly from contaminated soil to the human body (Bradley 1977, 7–15). The symptoms of *chalè* can be biomedically interpreted to include diseases in all five categorizations as well as conditions not related to water supplies. Stomach pains and digestive disorders can be caused by hunger, typhoid, and parasites. The condition called *glè* resembles a symptom of both amoebic dysentery and giardia (Jelliffe and Jelliffe 1985, 122–23). Hunger is also a cause of headaches, as are malaria and respiratory infections. Rashes on the skin can be attributed to scabies, impetigo, fungal infections, and eczema. There is also a condition called "prickly heat," which causes a red, itchy rash. All of the skin ailments mentioned here are linked to personal hygiene and hot temperatures (Harland 1985, 184). In addition, a burning sensation in the genital area is symptomatic of many venereal diseases. Dark urine, as well as many of the above-mentioned symptoms can also be caused or exacerbated by dehydration, an all-too-common condition in Haiti. Finally, one must consider the wide range of effects on the body that result from malnutrition, as well as the symptoms of heatstroke, which may include headaches, dizziness, and fainting.

The Nurses' Views

The two nurses in the town clinic each see over a hundred patients a week from the town of Bèlans and its surrounding areas. *Mis* Leyoni is the head nurse, and has been working at the clinic since it opened in 1995.[11] She told me that, on average, 60 patients visited the clinic each day, and, of those, 10 would claim to have *chalè*. She herself denied the existence of the illness, saying, "They come and tell me that they have *chalè*. But *chalè* isn't really a sickness; it doesn't exist as a sickness." She added, "I can tell you that it's because the people aren't informed. They don't have information. If the people were trained, if they were informed . . ." She compared *chalè* to *pèdisyon*, a condition among rural Haitians in which a pregnant woman's fetus is thought to shrivel up and remain in her uterus in a state of suspended growth and can be born several years later. In her mind, *pèdisyon* is also a result of ignorance. "We know that it never happens! Either the person is pregnant or she isn't! What happens is that the person may have a problem, or they've performed a strenuous activity, or an infection can cause that [miscarriage]. They say that it is *pèdisyon* because they haven't learned that that word does not exist!" In the case of *chalè*, she attributed its symptoms to parasites, acidity in the stomach, or high blood pressure. Surprisingly, despite her firm denial of *chalè* as an illness, *Mis* Leyoni explicitly stated that heat could negatively affect a person's health. She believed that heat could cause people to have difficulty breathing, become agitated (causing blood pressure to rise), and have pains in their head.

Mis Jinyoz, the other nurse in the clinic, was less willing to attribute any deleterious effects to heat. She also felt that *chalè* was a condition a person would believe in because he didn't know any better. When asked how she would react to someone who came into the clinic complaining of *chalè*, she said,

> I would consider that person to be a profane person, so he may feel that his head is hot, but it's really an organic problem that he has, such as anemia or vitamin deficiency. He can use the expression "*chalè* in the head," but it's probably because something is missing. People who talk about *chalè* in the stomach probably have parasites, that sort of thing, or gas. The people feel that their stomach is hot, but it has nothing to do with heat, it's just a problem that they have and that's how they interpret it.

She dismisses heat as a simple fact of life in a tropical country. "Well naturally, this is a hot country. Sometimes it's not, but it's a country that's naturally warm. But it's not true when they say that heat kills people. Sometimes the temperatures do get very hot, but when that happens, it doesn't get hot enough to kill people. When it's hot, you feel normal heat. Some people might feel irritated when it's hot." She did concede, however, that there were more cases of *kriz* during the summer months. The term "kriz" refers to a condition that can involve agitation, hysteria, and fainting.

The nurses' view of *chalè* has been influenced by their biomedical training and their social position, both of which foster a condescending view of the people they are treating. Both nurses acknowledged that older people were more likely to complain of *chalè*, with the insinuation that the latter were more old-fashioned and ignorant of modern science. Brodwin has written about the ways in which Haitian nurses publicly shame their patients, and I frequently observed this behavior in Bèlans (Brodwin 1997, 82). It should also be noted that both nurses have a much higher standard of living than most of the people they treat, especially in terms of income, housing, and diet.

Despite their refusal to recognize *chalè* as an illness, both nurses approved of the traditional remedies that patients usually use before coming to the clinic.[12] Both nurses also recognize, however minimally, the existence of heat as a factor that can influence health. As opposed to a condition like *pèdisyon*, which is completely impossible within the context of biomedicine, patients coming into the clinic complaining of *chalè* will not necessarily undergo the same repudiation of their beliefs. Because heat originates outside the body, it can be collectively experienced and objectively recognized as being "real."

Chalè and Larger Issues

Drought, Labor, and Poverty

Early in my interviewing process, it became apparent that there were larger issues associated with *chalè* than the illness as experienced by individuals. The exchange with which I opened this study (which took place when I asked a middle-aged farmer, "Does heat make people sick?" and received the response, "Yes, there's no water.") exemplifies the way informants moved the focus of my questions from embodied states to large-scale problems. Three major themes emerged: lack of water and land degradation, arduous labor and difficult working conditions, and poverty. An examination of how these topics appear in Haitian discourse on *chalè* gives a clearer picture of the illness, and provides insight into the struggles and suffering that are routine in rural Haiti.

Lack of water and land degradation are thought by many to be the most severe of Haiti's many problems. "Nothing better symbolized the vicious cycle of poverty in Haiti than the process of deforestation. Haiti was once a lush tropical island, replete with pines and broadleaf trees; however, by 1988 only about 2% of the country had tree cover" (Seyler 1991, 300). Charcoal is the only source of energy available for most rural Haitians, and countless acres of land have been cleared to provide fuel. Because the country is extremely mountainous, erosion is quick and severe. "Aerial photographs of the Haitian coast show huge buildups of sediment at the mouths of every river, evidence of the loss of topsoil in the interior. Haiti's topsoil is in the sea" (Wilentz 1989, 245). As the land becomes less and less fertile, farmers are forced to move their cultivation to more and more remote areas of land, further perpetuating the erosion cycle. In addition, sporadic rainfall causes severe flooding followed by long dry periods. Enduring drought has become a way of life for most Haitians. Many rivers have completely dried up and those that remain are often muddy trickles.

At first sight, Bèlans and its surrounding area present a slightly different picture. The hills around the town appear lush and green, and, compared to other parts of the country, the area has a substantial amount of vegetation. One reason for this difference is the lack of roads in Haiti's northern regions; few roads make it difficult for trucks to transport out wood for fuel. At night, however, one can see fires glowing in the hills as harvested trees are burned to make charcoal to sell in the immediate area. All of the land surrounding Bèlans is cultivated, with virtually no original growth left. Patches of bare land are beginning to appear, and there is little in terms of low growth—such as bushes, vines, or grasses—that can retain soil. The area has been suffering from severe drought. During my first summer there, it was reported that several people in a nearby mountain village had died of thirst. One man told me, "Even the animals don't find anything to drink. So many people are in drought."

It is not surprising that scarcity of water was a major theme in discussing *chalè*, given water's categorization as a cooling element. Lack of water makes it impossible to bathe, one of the main ways of treating the illness. In addition, the resulting effects on personal hygiene increase one's susceptibility to *chalè*. Even the small amount needed to pour over one's head or brew a herbal tea can make a significant dent in a household's water supply. When talking about scarcity of water, informants moved from speaking about embodied illness to the area's drought with no transition. One informant told me, "In the summers, there's a lot of sun, there's drought, and during those times, there's a lot of fever. That's when people get fevers. So it's an epidemic. This year, it's the same thing. Since June and July, there's a lot of sun and only a little rain. And you see so many people with fevers, so many in the clinic. Myself, I've just had a fever. So heat, it makes huge messes." In addition, heat combined with drought creates dust, which many informants identified as a carrier of disease. Finally, drought created a situation in which people had to walk further to obtain water and spend more time and energy carrying it back home. One man said, "When there's no water in the pipes, you have to go fetch water. All those people with their jugs, sun or no sun, people who are sick, they are forced to go. The sun pulls up the waters, so you have to travel through desert. Heat brings much, much, much difficulty." The public spigots in the center of town usually dry up during the summer months, and people (usually women and children) must get water either from wells in private homes or walk to springs in the hills, which can be up to an hour away.

Carrying water is only one of the many strenuous activities in which the people of Bèlans engage as part of their daily routines. In a town where few have electricity and even fewer have cars, human bodies are the main source of power, whether it be used to cut down trees, wash clothes, or simply walk from one place to another. As previously mentioned, physical labor and movement are factors that exacerbate and cause *chalè*. The illness is closely linked to hard work. During the interviews, it was interesting to note which activities and occupations were considered most likely to cause people to become ill. Farmers were consistently identified as the most vulnerable to *chalè*, because their days consist of climbing steep hills, digging, planting, harvesting, and other strenuous activities. Construction workers were also identified as particularly vulnerable to *chalè*. There are often small-scale construction or renovation projects underway in Bèlans, and work for them includes mixing cement, carrying rocks, and making cinder blocks, usually without the protection of shade. These two occupations are predominantly male, but women were identified as vulnerable when they were washing clothes at the river or cooking in a hot kitchen. In addition, the majority of market vendors in Bèlans are women, and, although a few have set up stalls that allow them to sit in the shade, most sit on the ground and are exposed to the blazing sun for hours at a time, day after day. Those pointed out as not likely to get *chalè* were people who worked in offices or indoor shops, although the latter were thought to be at risk if they were constantly lifting and moving merchandise. In general, there is an association between *chalè* and those who must work with their bodies for a living, as well as those who work outdoors.

Many informants drew explicit links between heat, *chalè*, and poverty. The following exchange exemplifies the connection that exists in the minds of many people:

Author: So you felt better today, because it hasn't been so hot out?

Informant: Yes, it's not too hot out. But later on, in the afternoon, around one or two o'clock, I don't know what will happen. It's about money. Money gives you more, you can see the doctor, there's relief. Economically, we're crushed. So heat, many people don't do well in heat.

Even when it was not explained in detail, it seemed taken for granted that more money would bring about a reduction of heat's ill effects. *Chalè* is associated with poor living conditions. One man spoke of his house, which he described as being so "poorly constructed, with only two windows" that the heat was unbearable at night and prevented anyone in his family from sleeping well. A mother holding her sick child told me that the baby had difficulty breathing at night, and that houses with low ceilings "provoked" *chalè*. Houses in Bèlans are generally constructed out of cinderblock and cement. Doors are generally left open, and there are no glass panes or screens in the windows. Homes are usually crowded. A man told me, "If you sleep in a house, in a room that has a lot of people, that produces *chalè*." The man with the poorly constructed house had said earlier, "*Chalè* affects people where electric energy doesn't work, and they can't find any economic means of fighting it. That's when it goes up." He added, "If our home had electricity, we could use a fan. It would give us a little protection." Another woman asserted that the type of iron she used (which was filled with hot coals) is more likely to cause *chalè* than an electric iron. Although the many discomforts caused by the heat and poverty in Bèlans are immediately obvious, it is only by talking to the people that one learns that they are also considered to be sources of illness.

Chalè is recognized as a problem that affects the entire population of Bèlans. Nevertheless, while speaking to some of the town's wealthier residents, I noticed some marked differences in their perceptions of the illness. The information given by two young women gives insight into how one's standard of living may influence attitudes toward *chalè*. The first, Antwanèt, was 20 years old at the time of the interview. Her family owns two stores, a large home, and employs a domestic servant. She is one of the few students in Bèlans who attend high school and is also a leader of her church youth group. When I asked her if heat made people sick, she answered,

> In Haiti, heat doesn't usually make people sick because the temperatures are not that high. Really, it's normal. It's not the same as in that place where I heard that a thousand people died, China, I think, where they died of heat. I don't know exactly when, but I heard it on the news. People died because the temperature was so high. And in New York, for example, it's very hot right now, so it's not the same thing as in Haiti, where the heat is normal, there's sun, but it doesn't kill people. It's not serious.

Antwanèt's discourse is reminiscent of the statements made by clinic's nurse, *Mis* Jinyoz, who described the heat in Haiti as "natural" and its effect on people as "normal." These answers are in sharp contrast with those of people who speak of the constant torment of the oppressive heat and the misery it causes as it "beats" people. Antwanèt later conceded that heat could cause *chalè*, and she knew how to treat it but continued to de-emphasize the severity of heat in her country. Although Antwanèt lived in the center of town and experienced many of the same discomforts as those around her, it is unlikely that she walked great distances, washed clothes at the river, or carried out any other activities that would expose her to harsh sunlight or prolonged, strenuous labor.

Another young woman, Klodèt, spoke in terms that more closely resembled those of the other informants. Although she had nearly completed high school in the city of

Okap and lived in a spacious two-story house, she also spoke of the heat's seriousness and deleterious effects on health. One of her comments, though, was particularly striking in relation to standard of living. When I asked her about people who worked outdoors getting *chalè*, she answered, "Yes, those people get *chalè*. But you'll find people who have worked their bodies so much, they're so used to work that they don't get *chalè*. People in the countryside, in the hills, who work in the sun." Thus emerge two contradictory interpretations: Klodèt and a few others suggested that one could become less susceptible to *chalè* by exposure to heat and work, with the idea that the body builds up physical resistance.[13] The majority of my informants, however, tended to emphasize that they were being worn down after so much labor, and that *chalè* was particularly harsh on their overworked bodies.

Conclusion

Folk Illnesses, Metaphors, and Culture-Bound Syndromes

At one level, *chalè* can be interpreted as a "folk illness," in that it lies outside the realm of biomedical diagnoses and has a distinct set of causes, symptoms, and cures. The term is limited in its applicability to *chalè*, given that "folk-defined causes are generally in the psychological/social realm, and, not surprisingly, have treatments that have psychological components" (Baer and Bustillo 1998, 246–247). *Chalè*'s symptoms are overwhelmingly physiological in nature, and psychological aspects were not emphasized by any of my informants. In their research on a folk illness called *caída de mollera* among Mexican-American infants, Baer and Bustillo argue that "folk-illness research has focused too much on asking people how they cure the illnesses, and too little on investigating how they treat the physiological symptoms of the illnesses" (1998, 242). In addition to recognizing the physiological components of folk illnesses and treatments, the example of *chalè* in Bèlans underscores the importance of studying their physiological causes as well, especially when they are related to people's living environments. Attention to these causes would not make obsolete the recognition of social origins of sickness and the cultural processes surrounding illness. Instead, the latter would be strengthened by a more informed position on the physical contexts in which ill people live. There are other ways in which the term "folk illness" proves inadequate in describing a condition such as *chalè*. Although the adjective "folk" can be interpreted to mean simply that the condition lies outside of biomedical classifications, there is an underlying suggestion that this type of condition represents a certain ignorance on the part of the people who experience it. As *Mis* Leyoni put it, "it's because the people aren't informed." Although the people of Bèlans have limited access to formalized education, it is condescending to brush aside the ways in which they give meaning to their experience and illness. The next logical step, of course, would be deny that anything is being experienced at all, as do some of the town's wealthier residents. Until the term "folk illness" loses its connotation of being fictional or imagined, it will be an inappropriate categorization for *chalè*.

This raises the issue of whether *chalè* is a "real" condition. It is possible that some people in Bèlans used the term *chalè* to describe ailments that they cannot explain. *Chalè* therefore becomes a metaphor, and, like many metaphors, there is no way it can be objectively studied and determined to be "real." In addition to the metaphorical uses, however, most of my informants believed that *chalè* is caused by heat entering their bodies, heat that cannot be considered a metaphor. Its effects are objectively measurable in terms of the low level of the river, the dryness of the land, and the undeniably high temperatures of summers in Haiti, from which there are few escapes. By dismissing *chalè* as a metaphor, one risks dismissing with it the lived experiences of a people who are suffering the effects of a prolonged environmental disaster from which there currently seems to be little chance of escape.

Finally, the term "culture-bound syndrome" has been used to describe conditions that lie outside the realm of biomedical diagnosis. One scholar defines these as "a variety of syndromes which, because of their seeming uniqueness and exclusive fit to local cultural conditions, are described as 'culture bound' " (Hahn 1985, 165). Two culture-bound syndromes that have received significant attention are *koro*, which occurs among ethnic Han Chinese, and *latah*, which is found primarily among women in Indonesia and Malaysia. Anthropologists have identified what they perceive to be social causes in both conditions. *Koro*, an illness that causes a person to believe that their genitals are shrinking, is most often seen among single, poorly educated men who believe in the condition and live in a social environment that provides them with little information about sex and sexuality (Cheung 1996, 79). *Latah*, unusual and often obscene behavior among people who have been startled, is thought to originate in Indonesian notions of propriety and sexuality (Kenny 1978, 210–211, 222). In these and other ethnographies, seemingly inexplicable conditions become clear once they have been put into their proper social contexts.

Looking for social origins of *chalè* will lead the investigator beyond the rural Haitians' discourse. Informants simply do not make links between social interactions and *chalè*. The links that they do make are more immediate and have been omitted from anthropological analyses of the illness experience. *Chalè* is an illness condition that emerges from and remains a part of a person's interaction with his physical environment. Scholars have pointed out the importance of recognizing the links between environmental factors and disease, but have neglected those between the environment and illness, or the human experience of being unwell. In pointing out that "agents occurring naturally or as a result of human intervention in the external environment may cause disease . . . unusual temperatures, electronic hazards as well as trauma, may produce pathology" (Sheldon 1990, 210), anthropologists have neglected to point out how *illness experiences* are created in reaction to the consequences of environmental degradation. It seems clear that most of the symptoms of *chalè* (stomach pains, skin rashes, and headaches) cannot be biomedically defined as pathologies directly caused by overexposure to sun, excessive labor, or lack of rain, yet these are the central factors in the illness experiences of the people in Bèlans.

Disease (transmission of pathogens)
(biomedically recognized pathologies)

Environmental Causes
(pollutants, new micro-organisms)
Social Causes
Illness (stressors pressures)
(human experience of suffering)
Environmental Causes
(land degradation, drought, poverty)

The illness *chalè* offers many opportunities to reexamine current ways of thinking about bodies, illness, and the environment. Theorists have pointed out that medical anthropology has "looked at issues such as bodily representation, sexuality and illness from a political economy of suffering perspective, which has tended to produce an incomplete portrait of the body (Green 1998, 4)." *Chalè*, with its clear links to physical factors beyond the political economy, provides the opportunity to begin mapping out a more complete portrait of the body, and the space in which it resides. For example, the illness calls into question the idea of the body as a bordered entity. If heat can strike a person's stomach in the same way that it can strike a river, what does this say about the body's permeability? Does this notion represent a type of relation between bodies and the elements that lie beyond people's skins?

The links that the people of Bèlans create between human bodies and environmental phenomena stem from an intimate relationship with an environment on which they depend for their most basic needs. In industrialized areas, this relationship is often tangled in a web of plumbing, highways, and electric wires. There is little room for romanticizing Haitians' links to their land, given their daily battles with overcrowding, deforestation, and irreversible erosion. Everyday realities grow harsher, while solutions remain scarce, making it clear that drastic measures are necessary. Rural Haitians will not be able to solve their problems simply through hard work and a positive attitude, both of which are currently being carried out with few successes. Without serious and far-reaching structural reforms, Haitians will continue to be subject to the economic and political forces that have led to their current plight.

If a person with *chalè* affecting every part of his body were to be admitted to a U.S. hospital, it is likely that he would be released that same day. He would be prescribed some antibiotics, ointments, and instructed to get adequate food and rest. *Chalè* conveys its full and troubling meanings only when it is put in the context in which it occurs. A man in Bèlans told me, "Children are really affected by *chalè*. They are unable to fight it, because they don't have much water in them." Haiti will only be able to begin healing when it has enough water to refresh both its land and its people.

Notes

1. The names of all individuals, organizations, and towns mentioned in this article have been changed.
2. This information is somewhat questionable given that the last actual census was conducted in 1982, and the 1995 report was simply an assessment based on the earlier figures.

3. Although I most often heard the words *chalè* and *cho* used in terms of heat or heat illness, these terms and related words are also used to describe other phenomena. Karen Brown, in her extensive research on Vodou practice in Haiti and the United States, has found that the terms *chalè* (to heat) and *echofe* (to heat or reheat) are used in Vodou to describe the creation of an atmosphere that will be conducive for shifts, changes, and transformations (Karen McCarthy Brown, *Mama Lola: A Vodou Priestess in Brooklyn* [Berkeley: University of California Press, 1991], 134–135). In ceremonies, creating this atmosphere is accomplished mainly by drumming, singing, and dancing. Such usage can be interpreted more than in a purely metaphoric sense: movement increases physical temperature within the worship space, usually a small, crowded room. I also heard the term used in this sense in Bèlans at a Pentecostal worship service, when a woman mentioned that it had been a long time since they had *echofe* in that church. She was referring to the ecstatic drumming and spinning that often lead to possession by the Holy Spirit, and speaking in tongues, which is characteristic of her particular congregation. I was also told that when a person is hot about something, it refers to a particular enthusiasm or energy. One man told me that heat referred to sexual desire. In general, there is a link between heat and activity/energy, which also holds true for the illness *chalè*.
4. I was unable to find information on hot/cold classifications in Central and Western Africa. However, given the widespread use of such classifications by medical systems around the world (i.e., Chinese, Ayurvedic), it would be a mistake to discount the possibility of African influences on Haitian humoral classification, particularly if one considers the overwhelmingly African roots of Haitian culture in general.
5. A condition that appeared frequently in my informants' discussion of heat and *chalè* was called *refwadisman* (cooling or chilling). A person gets *refwadisman* when he moves too quickly from a hot to a cold environment without giving the body time to adjust to the change in temperature. The example most often given was of someone going straight from a hot kitchen directly into the river to bathe. *Refwadisman* causes discomfort and aches and, although not considered life-threatening, was identified as a fairly serious condition. The prevalence of *refwadisman* in Bèlans testifies to the marked presence of a larger Latin-American humoral tradition, which emphasizes the balance and stability of hot and cold elements as a way of maintaining health. I was curious to know if moving from a cold place to a hot one could also cause sickness. Whereas some people said that it would not be a good idea to do so, most stated that sickness occurred only when going from hot to cold.
6. Although the description of *pouse* resembles that of constipation, informants described it with phrases such as "It's as if you're constipated," and differentiated the two conditions.
7. Few informants spoke of simply drinking water as a way of treating *chalè*. One young woman even warned that a person with the illness should be careful not to drink, "any old water," by which she meant water that was visibly unclean or of questionable quality.
8. One man told me, however, that drinking water after eating hot corn would give one a condition called *gonfleman*, or swelling. This is related to the rapid hot-cold transition described above in *refwadisman*.
9. This man defined infection (*enfèksyon*) as a lethal condition that could attack a person's bones, throat or skin. When I asked this same man what could protect children from *chalè*, he made reference to white blood cells of unequal strength, which were visible "in a laboratory," but did not elaborate.
10. The nurse, however, stated that hot weather agitated patients, which caused the rise in blood pressure, whereas the other informants linked it directly to the illness *chalè*.
11. *Mis* is the commonly used Haitian Creole word for "nurse."

12. There is an awareness in Bèlans of the value of medicinal leaf use even in the context of biomedicine. Mis Leyoni, when I asked how she felt about patients making teas to treat *chalè*, said, "Oh yes, it helps them. You know that all medicines are made from a base of leaves." Another woman in the town told me that she had heard that biomedical drugs were simply leaves that had been processed in a laboratory.
13. These ideas have a long tradition in Haiti where slaves and darker-skinned individuals were considered heartier and more resistant to work and the elements. It sadly persists today: one young girl told me that I should not fetch my own water or lift anything heavy because "whites" become sick if they work too hard or spend time in the sun.

Bibliography

Baer, Roberta D., and Marta Bustillo. 1998. Caida de Mollera Among Children of Mexican Migrant Workers: Implications of the Study of Folk Illnesses. *Medical Anthropology Quarterly* 12 (2): 241–249.

Bradley, D. J. 1977. Health Aspects of Water Supplies in Tropical Countries. In *Water, Wastes and Health in Hot Climates*, ed. McGarry Feachem and Mara, 3–17. New York: John Wiley and Sons.

Brodwin, Paul. 1997. Politics, Practical Logic and Primary Health Care in Rural Haiti. *Medical Anthropology Quarterly* 11 (1): 69–88.

Brown, Karen McCarthy. 1991. *Mama Lola: A Vodou Priestess in Brooklyn*. Berkeley: University of California Press.

Brown, Peter J., and Marcia C. Inhorn. 1990. Disease, Ecology and Human Behavior. In *Medical Anthropology: Contemporary Theory and Method*, ed. Thomas M. Johnson and Carolyn F. Sargent, 187–214. New York: Prager.

Cheung, Shueng-Tak. 1996. A Critical Review of Chinese Koro. *Culture, Medicine and Psychiatry*, 20: 67–82.

DeSantis, L., and J. T. Thomas. 1990. The Immigrant Haitian Mother: Transcultural Nursing Perspective on Preventive Health Care for Children. *Journal of Transcultural Nursing* 2(1): 2–15.

Farmer, Paul. 1994. *AIDS and Accusation*. Berkeley: University of California Press.

Foster, George M. 1994. *Hippocrates' Latin American Legacy: Humoral Medicine in the New World*. Langhorne: Gordon and Breach.

Green, Linda. 1998. Lived Lives and Social Suffering: Problems and Concerns in Medical Anthropology. *Medical Anthropology Quarterly*, 12 (1): 3– 7.

Hahn, Robert. 1985. Culture-Bound Syndromes Unbound. *Social Science and Medicine* 21 (2): 165–171.

Harland, Erasmus. 1985. Skin Diseases. In *Child Health in the Tropics*, ed. D. Jelliffe, 183–186. London: Edward Arnold.

Jellife, Derrick B., and E. F. Patrice Jelliffe. 1985. Parasites. In *Child Health in the Tropics*, ed. Jelliffe D, 110–123. London: Edward Arnold.

Kenny, Michael G. 1978. Latah: The Symbolism of a Putative Mental Disorder. *Culture, Medicine and Psychiatry*, 2: 209–31.

Kleinman, Arthur. 1988. *The Illness Narratives*. New York: Basic Books.

Laguerre, Michel S. 1981. Haitian-Americans. In *Ethnicity and Medical Care*, ed. A. Harwood, 172–210. Cambridge: Harvard University Press.

République d'Haïti, Département du Nord. 1995. *Population totale par sexe estimée en 1995 au niveau des differentes unités géographiques*.

Seyler, Daniel I. 1991. Haiti: The Economy. In *The Dominican Republic and Haiti: Country Studies*, ed. R. Haggerty. Washington, DC: Federal Research Division, Library of Congress.

Sheldon. 1990. In *Medical Anthropology: Contemporary Theory and Method*, ed. Thomas M. Johnson and Carolyn F. Sargent, 210. New York: Praeger.

Wiese, H. Jean C. 1976. Maternal Nutrition and Food Behavior in Haiti. *Human Organization*, 35 (2): 193–200.

Wilentz, Amy. 1989. *The Rainy Season: Haiti Since Duvalier*. New York: Touchstone.

Chapter 10

How *Houngans* Use the Light from Distant Stars

LeGrace Benson

Qur'an and Bible traveled on the ships bound from Africa to St.-Domingue. So did tracts of the European Enlightenment; copies of the *Declaration of the Rights of Man*; abolitionist letters from the Abbot Grégoire; Christian catechisms; *grimoires* full of instructions for magic rituals to insure love, good fortune, and victory. This print material was spare and simple compared to the complex intellectual environment carried around the minds of the people on board.

Some of the African captives chained below decks had in their minds the Qur'an, how to read and write in Arabic, long passages of laws and poetry, the methods for mathematically generating divinations, and the correct procedures for creating protective amulets—*hatumere*. Some of the sailors, especially the Bretons, had in their minds Latin prayers, Breton didactic exhortations and images, and a range of talismanic notions related to the severities of their profession and their desires for love and money. Desires for money even more than love seems to have been foremost in the minds of the chief officers of the ship, to judge by their actions and some of their reports. And there were the priests, their documents revealing minds full of devotion, cruelty, incomprehension, dissolution, courage, sanctity, erudition, and ignorance.

The names of some voyagers appear on ships' manifests: the captain and first mate, some of the seamen, the plantation overseer, the priest come to baptize the Africans with new names in the name of the Father, the Son, and the Holy Spirit. Once on the plantation, some amanuensis inscribed the imposed names in fine French script on the habitation *inventaire du biens* (inventory of goods): "Noël," "Minerve"; Christian words hiding the Mandinge name "Noël" knew to be his own, the Kongo name Minerve received from her parents. "Noël" was worth 2,200 livres to the Chavannes habitation, "Minerve" only 1,600.[1] Who knows what they were worth to their companions? The inventories left aside any accounting for wisdom, knowledge, and the techniques of the sacred.

On the ships, in the streets and markets of the ports, and on the plantations, the people caught up in the traffic of human beings and raw products spoke with one another, their pidgins and rapidly developing creoles, a linguistic triangulation between owner, owned, and the daily conditions of life encountered by both. Their several enlightenments necessarily had to be shared. Had there already been, on board, little collusions between sailors and the cargo of captives? Both were wretched of the earth, endangered by the sea, disease, hunger, thirst; indentured or enslaved, their human conditions comparable though not equal. Did the sailors look the other way if some captive had a precious burden of Qur'an? Some of the sailors were Muslims passing as Christians: they surely would have protected the sacred objects. It is certain that once in St.-Domingue portside, the books of magic that the Breton sailors carried in their duffels went into circulation, quickly finding their way even out into the forests with the maroons. What use might the sailors in turn have made of the strong protective invocations and amulets of the Muslims? The fashioning of a new religion using materials from African traditions, African and European forms of Christianity, African Islam, Breton seafarers' folklore and magic, vernacular Catholicism, each luminous stand borrowing and giving light to the others, began on the boats and in the ports. The New World *houngans* were cosmopolitans.

Material evidence for the braiding of so many strands is scant for the first century and a half. Tropical weather and anti-superstition campaigns destroyed much. Only toward the middle of the twentieth century when Haitian Indigenists called for new valorizations of African heritages and of rural wisdom did there begin to be the deliberate photographic documentation, oral history, and preservations that allow firm assertions. Yet certain patterns of graphic design, certain reiterated behavior elicit queries important enough to merit study even in the absence of "answers."

There are questions of continuity and restoration, in part a matter of narrative strings and technical skills passed on from one person to the next, maybe as far back as a *bossal* (African-born) fresh off the boat from Ibo-land, or as recent as the new insight a *manbo* (priestess) passes on to her granddaughter. The formation of habits of attention and habits of behavior results from Long Conversation in actual place and time between and among specific individuals. Rather than being a non-observable "collective memory" or "collective unconscious" operating in an elusive genetic *etherium*, the exchanges take place between namable people and are a major means by which they continue and transfer their degree and style of illumination.

To call this a "Long Conversation" names only some of the processes taking place. "Long Theater" encompasses the actions and ideas more generously, but obscures the intimacy with which most such transfers occur. While carrying on "conversation/theater" is absolutely crucial, the fundamental conditions upon which this must rest are beyond as well as beneath the show-and-tell. The ground is individual apprehension of the environment with all its possibilities and impossibilities for human behavior. The conversation/theater is social. Yet, whatsoever habituations of attention and behavior result from shaping of the community present and historical by the Long Conversation, the fundamental material—the ground stuff—comes from singular persons, each one of whom needs to know and therefore discovers where to walk safely, where to find shelter, what is safe and good to eat, where the water is, where to rest, and where to find love and be powerful. Seeking these out is accomplished by using senses and sensibilities; once found, the knowledge, wisdom, or skill is secured for the single person *and* companions with the Long Conversations. *Habits* of attention are generated in the Conversation/Theater; attention itself is a perceptual function accomplished by one person at a time. Objects and actions of Vodou demonstrate the reciprocity established between the singular and the communal, creativity and continuation of the Long Conversation/ Theater.

The captives who were respected imams, Christian clergy, Yoruba or Arada or Kongo *houngans* and *manbos* were exactly those members of a community or kingdom who were most alert to the actual purities and dangers of a particular environment. Their teachings, prayers, healings, and protections had to work satisfactorily in the daily world. They had to be continuously attentive to material, plant, animal, and human conditions and changes. The fundamental, ground content of their operations rested as much on astute gathering of information as on their spiritual comprehensions and received knowledge. The style of presentation and manner of operating on behalf of a congregation or an individual petitioner might show the shaping marks of the Long Conversation/Theater current in that particular locality and among a particular people, but the traditional material and style had to be in resonant accord with the immediate, real needs of a congregation. Metaphors and figures of speech that are efficacious poetically and religiously are tied and bound to real affordances for real behavior. The *houngans* know that. The *drapo* for Ezuli is waved about for real love; that of Ogou for real victory. The *houngans* and *manbos* know the clients and the environment well enough to tie each metaphoric emblem and gesture to real necessities. The imams, the *houngans*, the diviners, and seers may take a while to reacclimate to a new environment as they did arriving in St.-Domingue, but fundamental shrewdness and experience with exact and precise discriminations and detections of the environment stands them immediately in good stead. They will use what is familiar from old times and places, reintegrating into a present environment. Their companions will regard them as leaders and sages.

The Long Conversation and Light from Islam

Stories from all over the Americas tell of the influence of devout Muslims, especially those who were imams and marabouts.[2] By the time of slave trade Mende nations already played a major role in the spread of Islamic ideas, images, and practices throughout the elaborate trade systems of Africa. Where their trade flourished, so did their religion and their style. Local residents in areas where they had a trade and cultural hegemony are reported to have transformed histories and genealogies to show connections with these successful people.[3] LaBelle Prussin has described how West African centers of learning and commerce, such as Tambuctou, Djenné, and Kongo, from thence south and west to the coasts, impressed their symbols, their ways of ordering and arranging both time and space, their religious and domestic architecture, and their religious practices upon the underlying local governances, religions, and everyday objects and habits.[4]

In Africa the Islamic Long Conversation coexisted with the cultures predating it. West African history from the eighth century forward is one of the rise and fall and rise again of Islamic hegemony in contest with or in assimilation with local ways.[5] The university city, Tambuctou, was a commercial nexus as well, lying as it did in the geographically key point between sahel and savanna and on a river. Other important transfer points developed at Chichit to the west, Gao to the east, and Djenné, Bobo Dioulasso, and Kong to the south, a chief point of Islamic contact for Old Ghana. By the time of Portuguese incursions, Islam and the original cults were intimately interwoven. So were the populations and so were the contentions between rulers of one district and another; out-of-favor or war-captive Muslims were among the first slaves sent to St.-Domingue from Senegal. Captives seized for slavery included a number of impressively capable persons.

The role of imams and marabouts as advisors to rulers and as leaders of schools and universities gave them an aura of authority in Africa that would cling to them despite captivity. Notwithstanding their relatively small numbers and their cultural differences from adherents of either traditional African religions or Christianity, their presence in St.-Domingue as companions of the journey and as literate cosmopolitans with a well-formed cultural and religious identity would have graced them on the plantations with the prestige they held in Africa. Prestige was securely founded upon literacy, personal discipline, command of contact languages, and their rich arcana of numerology, astronomy, and mathematics; their insights into efficacious methods for divination and healing. Rudely separated from the precious Qur'ans, their instruction boards, and their mystical painted and embroidered fabrics, or having to hide the few things they had smuggled in, they seem to have tried to re-create them.

Especially in the eighteenth century, manifests and inventories show heavy importation of Mende peoples to Hispaniola, as is related by Fouchard and others. In *Marrons du Syllabaire* Fouchard notes observations as late as the twentieth century of a continuing Muslim presence. He cites the example of a door, now lost, in the region of Jacmel upon which Arabic signs and writing appear.[6] Thomas Madiou notes the influence of Islam during the early uprisings as well as in the precipitating events of the Haitian Revolution.[7] Contemporary historian Serge Fuertes attaches special significance to Malinke (Mende) influence.[8] J. R. Emmanuel Francius Julien makes interesting points in this regard, noting the prohibition of pork by Muslims as a sign that the sacrifice of a pig during the proto-revolutionary Bois Caiman ceremony was related to extreme breaking of tabu, thus indicative of the seriousness of the oath that the revolutionaries took.[9]

The imams had already been in Africa for centuries.[10] They arrived on the first and last slave ships and put their great knowledge to the needs of survival first, then of revolution—a *jihad*, one might claim.

These enslaved arrived in the New World centuries after the first Muslims to make that landfall. Abubakri of the royal house of Mali led an expedition to what is now called Brazil in 1312. Roger Bastide alludes to early Muslim contact in *The African Religions of Brazil*.[11] The Malian colonizers may have sent emissaries to Hispaniola before the fifteenth century, and early contacts between them and the Taino are a possibility.

The proto-revolutionary, Mackandal, whose name became synonymous with "poison for enemies," was not the last Muslim to cross the channel from Jamaica to

Haiti. From 1838 to 1924 planters in the British West Indies imported an estimated half-million indentured servants from India. David Lowenthal estimates one-sixth of those to have been Muslim.[12] While many of the indentured imports remained in British islands, a number moved into Haiti over the decades. Today near Cap Haitian, there is a recently built mosque, perhaps the result of the work of a missionizing group active in Jamaica since earlier in the twentieth century. This group may or may not be in close contact with another group nearby, who say they practice "Mandingue" service rather than Vodou, and are described by their neighbors as eating no pork and of reciting prayers each day at sunrise and sunset.[13]

With so many opportunities for making a presence felt, questions arise as to why the presence is attenuated and the evidence so discontinuous and elusive. Several currents pushed against the ship of "Imamou." Especially vigorous were those of the Reconquista, the anti-literacy campaigns that were waged from the beginning, the power of the ancestors and the ancestral African faiths, and the colonial environment of dangers, insecurities, and deprivation. These currents arrive with some of the same tides that brought the Bretons, whether petty merchants, seafarers, or missionaries, to St.-Domingue.

Reconquista

January of 1492 saw the final success of the reconquest of the Iberian Peninsula from the Muslims. The Spanish crown had a papal mandate to overcome the "infidels" wheresoever encountered. Pope Alexander VI specifically bound Christianity with the exploration and conquest of "America." Abdullah Hakim Quick outlines a sequence in which Moriscos, nominally converted to Christianity, reverted to Islam once in the new World, and actively sought to proselytize among the "Indians." They enjoyed enough success to motivate a series of anti-Muslim laws by King Ferdinand. His decrees did reserve that "Moors" were not to be taken as slaves.[14] Later, in the era of Pope Nicholas V, the French would apply the justification of religion to enslavement of the Barbary Muslims, then eventually to Muslims of any provenience.[15] John Thornton describes alliances between Europeans and West African politics, for example, that while primarily economic in gestation and aims, attached religious fervor as well.[16] Kongo kingdom was Christian before the Columbian expedition, and Elmina in Old Ghana had a church dedicated to Sant'Iago (St. James the Greater, patron of the Iberian Catholics) in 1480.[17] Christian slaves, especially those from Congo, enjoyed preference by missionaries as catechists to other slaves.

William B. Cohen describes French colonization and slavery enterprises as having much the same motivation as that of Spain's Ferdinand, adding that especially in response to events of the Counter Reformation in France "expansion was seen not only as the expansion of the king's domain and of the state's wealth, but also as a missionary effort. In 1603 the French King declared that his main aim in supporting expansion in the New World was the conversion of the Indians."[18] Ironically the "Indians" to be converted in St.-Domingue were either dead or hidden in maroon enclaves in the mountains. The missionaries sent from France would encounter

Muslim, Christian, and Traditional Religions adherents. European toleration of Muslims and Jews during the era of Reconquista extended only to those willing to convert, and only to about 1500. Subsequently, suppression and "purification" of Islam continues to the latter decades of the twentieth century. Muslims in St.-Domingue would find it impossible to observe any of the five pillars of the faith, except perhaps almsgiving in the form of sharing what little they had. The obligation to make ha'j, the pilgrimage to Mecca, was beyond even a symbolic journey. The tenacity of Islam in the face of such repressions is remarkable even as vestiges.

Anti-literacy in the Americas

Islam is missionizing and universalist, dependent upon pan-societal literacy. (In the West African lands, as in the time of Muhammed himself and immediately thereafter, literacy included women and girls.) In the Americas where the Muslims had to relinquish their books, they set about re-creating them from memory, painstakingly, for an error was an error against Allah.[19] Colonists saw literacy in any language as contrary to their interests. Some clergy and religious people in St.-Domingue supported literacy, but only the French and Latin of the Mass. To set up *ulaama*—Qur'an schools—would be exceptionally difficult. Moreover, benefits of literacy are long-range: slave conditions required immediate response to demands from the masters and the often terrible exigencies of the environment. Anti-literacy in fact became an enduring condition in St.-Domingue and in Haiti.

Human speech evolved as a genetic ground-condition; reading and writing as intellectual invention without any apparent genetic continuity. The mystical and esoteric qualities that get attached to writing probably derive in part from being based in a social contract. Literacy can by social contract be extended or withheld. In contrast, perceiving, acting, and speaking are universal competencies. Where literacy is proscribed, orality and theater become more fully exploited. So it was in St.-Domingue/Haiti. Slaves from cultures where orality and theater, including music and dance, predominated had competencies better suited to leadership in an anti-literate environment.

Symbols and representations likewise survive despite injunctions against reading and writing. In the case of the time-consuming and complex verbal and mathematical operations needed for a true *gemmatria* to generate an authentic *hatumere* or magic square, an imam or someone taking the place of the now absent imam could draw a representation, something that had the look of the magical device, and fill out the spirit and desire with hope and prayer. Complexity and depth evaporated, but the trail where they once were remains marked.

The Endurance of the Folk Beliefs and Practices

In the plantation environment hostile to literacy, spoken, sung, acted-out, and symbolized sensibilities came to the fore. Since needs served by the literate imams were more pressing than they had been in Africa, the Islamic experts continued to exercise their skills as seers, diviners, makers of protective amulets against dangers and witchcraft, healers of animals and people; continued to be active in religious and "magical" everyday life, developing a range of practical actions and symbols oriented more to everyday needs than to transcendencies. The healing and divination designs associated with them probably lingered on into the present because they were in good accord with Fon, Ibo, Congo, and other ancient traditional practices. Evidences of material culture from the plantation era enduring into the present reveal the probable power of the engagement.

Pierre Pluchon cites from the letter of Jesuit missionary to the French colonies, R. P. Mongin, a description of the sortileges of slaves who were called "marabouts," in which he particularly takes note of patterned belts used for protection against the pains from beatings.[20] Such belts continue to be made in western Africa to the present.[21] The pattern on the belts sometimes directly encodes names of Allah, sometimes only a graphic representing the positions of the names. The pattern appears in modern Haiti without the names and on objects other than belts. Further studies of Bizango societies (where they use binding chains, some of them claimed to be from the slave era) may turn up related belts or other tied or tying items showing this old Islamic pattern. Continued study of the society in the north of Haiti calling their ceremonies "Mandinge" may also yield artifacts with Islamic protective devices.[22]

The Power of the Ancestors

Reports on African religions from earliest European contacts to the present often characterize the adherents as "worshipping ancestors." Perhaps this error results from the contrast of regard for ancestors and recognition of their continuing presence with Christian practices of consignment of their dead, *incommunicado*, to an unknowable Beyond: Purgatory, Heaven, or Hell. Christian's beliefs concerning those "fallen asleep in Christ" are complex and variable, but in general, the notion of direct communication and daily contact in the dwelling, the yard or the marketplace is uncommon or even proscribed. Community engagement with the ancestors is thus striking to Christians, who denote it as "worship." Christianity and Islam, stressing the divinity of the One God, see heresy in attention to ancestors.

Yet, those Christians who venerated saints (who are, after all, ancestors); where they invoked their assistance for guidance, protection, and healing; and especially where there were special images, processions, and services in their honor, Christian habits of attention could easily be assimilated into the attentive veneration of the deceased family. The so-called worship of ancestors was practically and theologically compatible with the notion of the communion of saints embracing those living visibly or invisibly in the many mansions of Bon Dieu. The banners of the Christian saints could make the African ancestors flutter in the breeze, releasing doubled power.

Islam, less attuned to the harmonics of societies living with the presence of their ancestors, was less amenable to such assimilations. Instead of conversations with the dead, or invocations to the saints, the sacred, written word of Allah as transmitted to

Mohammed by Jabril was the ever present help in time of trouble. But the trauma of separations not only by death but also by captivity intensified a desire to re-create and rebind the absent back into one's habitual attention; the Surahs were insufficient comfort. On the other hand, slaves could be rounded up for Sunday Mass, where they would encounter the ancestor whose Body and Blood were consumed by the faithful. The Mass was integrated into the New World African religion, Vodou, early on and continues to be so. Allah was a Name of God and Sacred Word; Allah was Inconceivable Creator; not Brother and not Sacrifice.

The Power of Attention to Environmental Inspiritedness

Another major misperception of African religious practices on the part of Europeans was the notion that objects in the environment—trees, animals, rivers, thunder were "worshipped"; it is called "animism" to this day. Perceiving the world with its myriad transformations and metabolisms, traditional attitudes throughout sub-Saharan Africa recognized Being-existence as lively Presence in all things. With such a perception, each thing is understood as an actual or potential mediator or intermediator of forces and tensions of the environment. While sacredness of the created environment is adumbrated in Torah, Bible, and Qur'an, the West African traditions devote far greater attentiveness to the inspiritedness of water, tree and grain, beast, sun and storm. St. Francis of Asissi and Teilhard de Chardin among Christians are notable exceptions to the trend. Both Islam and Christendom would turn their attention to discerning order and causality as means of explanation, prediction, and control. This orientation of attention contrasts with that focused upon living in harmony with the forces present in the world.

In St.-Domingue/Haiti, the various styles of African sensibilities would have been resonant with Taino belief and practice. Maroons fleeing into forests would have found a new but compatible mode of attending to, understanding, and coexisting with the *an deyò* (back-country) environment. The oldest of the African habits of attention would have been a more readily useful and fortunate legacy in such circumstances.

Yet the Imams' Signs Survived

Despite all pressures to the contrary, the presence of the imams is still fleetingly visible. Writing as mystical power in itself, coupled with the power of the Names of Allah and inscriptions in magical diagrams of His Sacred Words, came to be appreciated as efficacious. The protective belt seen by the Jesuit exemplifies a common practice. Even if the belts and amulets themselves disappeared in tropical damp and the bonfires of the suppressions, the design devices persisted.

The Flagg Collection of Haitian Art of the Milwaukee Art Museum includes a painting by Valcin, "Ceremonie." It depicts a Vodou service taking place on what appears to be a tiled floor, which may not "mean" floor at all since contact with the earth is essential for the ceremonies. On the checkered tiles is a *vèvè*—one of the sacred signs drawn at the beginning of each Vodou ceremony. The grounding of this painting thus really has two sets of sacred symbols, one the tiling, the other the *vèvè*. Divinatory technique in West Africa involves such a checkered pattern: the painting alerts us to look for its source and implications.

An example far from Haiti can illustrate how such a drawn device implicates action in space and time. A flat, sand-painted mandala of Namgyal Monastery, extrapolated by Cornell computer programmers, reveals a dimensional sacred temple mountain, the site of particular spiritual events, and the stage setting for playing out the balance and harmonization of cosmic forces.[23] In Tibet, as in Haiti, divination, more than a means for telling a future, is a means of bringing spiritual and earthly forces into proper balance and harmony. Karen McCarthy Brown has extensively analyzed cosmic forces similarly encoded in *vèvè*.[24]

The Valcin painting and the mandala each exemplify signs used to point to vast, complicated events and to relationships between ordinary humans and transcendent forces observed to work on and through them. In the Haitian example, the checkered pattern is almost certainly more than just a stylish background. It looks like the familiar game board, which is in turn a destiny board: the magic squares of Islamic divination.[25] It is these squares that appear on the *drapo*—the Vodou ceremony banners. Like the game of chess from which they may ultimately descend, they are symbols for and ground of warfare. They are the layout for the Great Battle: literally, figuratively, and spiritually the battle of life against death; against that shadow of death, sickness; against death's other terrible shadow, failed love.

Senegalese and Malian imams or marabouts were among the first arrivals in St.-Domingue and may have been the first "peoples' priests" after those of the Taino. They functioned in Africa and presumably in Diaspora as teachers, advisors, healers, diviners, and makers of amulets, in each of these operations continuing their African practice of using designs from *al-shah mat*, the game of chess (*al-shah mat* = shah is dead; hence checkmate, the king is dead). The design was a checkmate against enemies. It is highly likely that the functions of the magic destiny squares, known well to the earliest slaves, became wholly subsumed into the fabric of Vodou. The checkered pattern along with certain other arcane Islamic signs, the *khawatim*, appear in many *vèvè* and (much) later in *drapo*. Such *drapo* perpetuate tacit knowledge (from the Long Conversation/Theater) of the destiny squares of the imams. To swing and ripple the banners within a sacred space, to wrap up in them as though in armor, is to pass their protective force against death and its two shadows into the bearers and into the congregation.

Anna Wexler in "The Artistry of a Haitian Vodou Flagmaker," describes the *houngan's* process of designing, executing, and using the *drapo* in ceremonies. Religiously devoted attention to detail secures the efficacy of the flags. The circumstances of conceiving, making, and using are strikingly similar to those reported of imams and marabouts of West Africa, making their talismans and divinations over matters of life and death. Wexler reports deep mediations and prayers before any work is begun. She relates, "As active presences during important ceremonies, the flags extend the vigilance of the houngan or manbo (female priest)."[26] Delores

Yonkers corroborates: "Carrying and caring for the expensive banners involves considerable knowledge and responsibility. They are believed to possess a magical power which can be activated in the movements of the ritual and must be revitalized by being stored furled in closed shrines, leaning upon the altar when not in use."[27] Reverence for the *drapo* is close to the Islamic belief that writing and embroidery in and of themselves hold spiritual power. Is this a continuation of a Long Conversation about proper action to insure efficacy? Unmistakably it shows a link between graphic symbol and physical actions so distinctive to Vodou.

Yonkers emphasizes that "the military tone of the use of the flags in vodoun ceremonies is implicit. When the paired figures appear so suddenly and so aggressively, they flank the sword of Ogun [who is also St.-Jacques Majeur]. . . Defender of the besieged, representing the ideals of justice and liberty, his representatives direct the movements of the entourage in a choreographed ballet of blade and fabric."[28] In "Sacred Banners and the Divine Calvary Charge," Patrick Polk calls attention to the strong resemblance of some *drapo* to banners of Napoleon's army, citing the latter as a probable source for many of the devices and emblems of the former.[29] While Yonkers and Polk both rightly emphasize the often bellicose strivings that impel participants in a ceremony, the war game of chess may move us deeper into both Napoleon's territory and the universe of Vodou discourse, informed by Islamic Long Conversations.

When the Arabs conquered Persia, they picked up the game of *chatrang* in which winning was *al-shah mat* (the king is dead). This phrase became Sheik mate, check mate, *échec*. In the centuries of their domination of North Africa, a period coinciding with the spread of Islam, Arabs carried the game along with their religion across the Mzab and Magreb, down to Tambuctou, and from there into Kong Goa, Djenné, perhaps eventually even as far as the Ghanaian port city Elmina—everywhere there was Muslim trade and hegemony. The Moors took their game into Iberia, *al-andalus*. There are drawings in the Al-hambra palace of people playing the game. The checkered board and its playing pieces came to appear on European battle flags and coats of arms, perhaps for some of the same reasons these emblems would later appear on Vodou flags. The magic square version begins to appear also in alchemy and other books of esoterica. The visual documentation available from the sahel and savanna and south indicates its use primarily for protection, healing, and assuring victory. Yet, in western Africa, where present evidence ties it to victories and healings, there is an old tradition of playing the game. In Ghana, today, men are reported to play chess "for hours, not even stopping to eat."[30] The Asofo regimental flags from the era when Ghana was a British colony show the checkered pattern in a number of interesting deployments, one of which has a British lion playing chess with a Ghanaian leopard, possibly related to a proverb expecting eventual victory for the leopard.[31] One of the Asofo banners in a traveling exhibition depicts *al-Burak*. In this case the Islamic influence is unmistakable, as this beautiful mare with the face and streaming hair of a woman is the steed who carried Muhammed to the Seventh Heaven in the Night Journey. She will appear in Haiti.

A Qur'an board now in the Brooklyn Museum[32] shows the checkered pattern flanked by battle tents, thus underscoring the association of the pattern with warfare, and by metaphoric extension, any agonic contest. It is reasonable to conjecture a more than coincidental parallel between such an image and St.-Jacques/Ogun caracoling his white charger over or within a checkered field. Such a *drapo*, devised out of mediations and dreams of a Vodou *houngan*, appears for certain in its shining sequined version only after 1960, and in urban areas. But there are much earlier examples of modest painted *drapo* that show the patterns of the magic squares and the *hatumere* amulet designs. Erika Bourguinon collected two such patterns during field work in rural Haiti in the late 1940s and in the 1950s.[33]

The chessboard layout was an excellent war game field and instrument for divination because the possible moves on the bicolored board are factorially large. In the medieval era the board and its pieces served to encode secret messages. The pattern and its variants, *lozengy* (diamond) and *fusilly* (elongated diamond) appeared on European war banners and on the costume of Harlequin. In Africa the three patterns became ubiquitous, often but not always associated with Islamicization.

Of the pattern and its variants J. E. Cirlot has written of how they express duality and contest of counterpoised dynamic forces. "It is significant," he writes, "that the costume of Harlequin (a chthonian deity) is actually made up of *lozenges*, which proves beyond a doubt that the harlequin is related to the gods of destiny."[34] This intriguing claim parallels the use of the pattern in, for example, the Kufulo subgroup of Islamicized Senufo of Côte d'Ivoire where the "harlequin" costume clothes dancers for the funeral of blacksmiths. The patron deity for blacksmiths in African lands as well as in St.-Domingue/Haiti is Ogun. In Haiti variations of the pattern appear on recent banners for Ogun/St.-Jacques, but much earlier on *vèvè* for the *lwa* (spirit) Azaka Médé. But we run ahead of the story.

Henry Gough and James Parker, writing in 1894, suggested that the checkered pattern might have come from a counting board, called a "chess," widely used in English inns and pubs.[35] This anachronistic supposition reveals a numerical function associated with the pattern from its beginnings. Islamic mystics, by a process called "gemmatria," allied numbers and letters to the pattern so that either could be added or aggregated in all directions, yielding the names of Allah. The gemmatria became further related to astrology and alchemy.

People throughout western Africa regarded both writing and calculation as mystical, portentous acts. Destiny, divined by means of these was tied to financial matters, to health of people and cattle, and to the fortunes and misfortunes of love. The board and its games and divinations had implications for the essential facts of life. Little wonder then that the pattern and its variants appeared and continue to appear in a wide range of protective and good fortune objects. The enduring regard for its power is evidenced in power signs currently in use in Africa and wheresoever Africans and their descendants are in diaspora. Maude Wahlman reports a quilt made by Arester Earl of Georgia, United States of America, called a "mojo" quilt. It consists of small squares in contrasting colors, aggregated in larger squares in turn aggregated into the square quilt.[36] She discusses mojo bags, often used as amulets, and containing various herbs and other substances to ward off evil or attract good fortune. This is similar to Azaka Médé's checkered bag and its contents. These would sometimes have a nine-piece mojo square painted on the outside, sometimes the outline of a hand. Eli Leon asserts that most block-style quilts of America relate back to African rather than European models.[37] Perhaps the ultimate African source is the magic squares/destiny game. The checkered pattern appeared frequently in British heraldic devices, yet seems to have been ignored by quilters there and in the American colonies. A possible reason for this lies in the accustomed use of the pattern on the one hand to denote families of the peerage, and on the other to invoke and control great forces impinging upon ordinary life. The parallels with aid, destiny, and even the healing herbs contained in the check-marked sack of Azaka is strongly suggestive of the actuality of a Long Conversation continuing throughout the African diaspora.

Susan Tselos notes that borders of squares and *lozenges* only begin to appear on Vodou *drapo* around the late 1960s at the same time sequins became plentifully available. She surmises that the attention given these borders was "possibly a reflection of the 'op art' and graphic textile designs so popular in the United States and Europe at the time."[38] Flags older than the 1930s are extremely rare, in part due to the ravages of the anti-superstitions campaigns, and Tselos documents the change in those remaining from no borders to small ones to those that are wide and richly colored. The object described by Drouin de Bercy in the late eighteenth century, which was a large cloth sprinkled with blood, feathers, and the talons of birds may have been an early version of *drapo*.[39] *Drapo* made after 1960 are "cooler" in their materials, and Tselos's comparison with contemporaneous garment industry chic is appropriate on several counts. This was a period when *drapo* began to be made for sale, not simply for ritual, and when market factors such as the tastes of tourists began to exert influence.[40] Yet some of the flags continued to be made by *houngans* for Vodou ceremonies, including some that later came up for sale in the art market. The patterns perhaps look like "op art" in the same way and for some of the same reasons the *lwa* Ezuli looks like Our Lady of Perpetual Help.

Let us catch sight again of Azaka's checkered straw sack filled with healing herbs to follow him back into an *an deyò* of time and place. In photographs made in the 1930s in rural (*an deyò*) districts by Courlander, Herskovits, Deren, and others, the magic squares, destiny squares, mojo squares—whatsoever they may be called—appear on the *vèvè* for Azaka Médé himself, Bossou, Ezuli, Ogou Badagris, Gran Bois, and Gedé the brother of Azaka.[41] Milo Rigaud in *Secrets of Voodoo* illustrates a quadrille heart of Ezuli, centered in a pentangle (seal of Solomon in both Islam and Freemasonry) within a square, the result looking much like a *hatumere*-Islamic amulet. Rigaud translates the incantation associated with it as "Sakpata = Impurity."[42] Sakpata is the spirit (some say healer) of smallpox encountered in western African lands. The Haitian version suggests yet another exchange in the Long Conversations. Unpublished designs photographed by Herskovits also show a *houngan* drawing a *vèvè* around a bier, in which *khawatim* appear, again designs of some antiquity, of Islamic provenience, and associated with life and death matters.[43]

In the post-United States Occupation of Haiti there was an energetic refocus of attention away from French heritage coupled with counterreaction to the perceived racism of the Marines. Educated urbanites began to pay attention to the *an deyò* African traditions of the rural people, "Ayiti Toma." Some ancient African patterns for drapo and other ritual objects survived out in the *an deyò*, and were at hand to be used. In western Africa there are war shields and weapons marked with "arabesques" that only appear to be writing, an inscription representing writing.[44]

Similarly, the squares and *lozenges* of Vodou *drapo* and *vèvè* could be representations of the quadrilles of the Qur'an boards and amulets: outward and visible signs to invoke an inward and spiritual presence. The *houngan-drapo* artist, Clotaire Bazile, said of his wide, multicolored borders surrounding Ogun/St.-Jacques, "Oh, we've always done it that way."[45] He went on to say that he dreamed the designs before he drew them out on cloth. Bazile's designs are distinctive, as are those of other skilled artist-*houngans*, yet they are drawn from old times, and in a process that is entirely comparable to the procedures used by the imams and marabouts to create an amulet or a divination. The harmonies of the *drapo* made in this way are themselves signs of intent of good destinies, to be understood as implying time and space related to a larger cosmology. The habits of attention are profound and ancient. In some ways this parallels glossolalia which sound like language but have no grammar or syntax, yet are understood to be inspirited in ways similar to the inspiritedness of scriptures or spoken ritual formulae. Someone thoroughly familiar with the *langaj* of Vodou ceremonies might listen attentively for Arabic or pseudo-Arabic vocables. "Pater noster in coeli . . ." or "Ave, Maria, gratia plena . . ." as well as the Aramaic phrase from the New Testament, "Lama sah-bha-k-than-ni," are reported. So are "ar-ch-allah" and "ku-m-bha-lah-dya."[46] "Allou akbar" is clearly sung in the Mandingo ceremony Rénald Clérismé recorded in the north of Haiti in 1997.[47]

Continuities from Islamic West Africa were probably first transposed from amulets such as the belt seen by Mongin, directly in the process of invention of certain of the *vèvè*. The seven signs of the *khawatim* may have moved from the amulets to the Vodou signs in the same way and at the same time. The *khawatim* "are mysterious marks said to represent the excellent names" or symbols of God. Only the first sign, the pentacle, appears to be understood: it is known as the seal of King Solomon.[48] The pentacle appears in some old photographs of *vèvè* as well as on modern *drapo*. Djinka cloths discussed in several books on African Islamic art, including those by Bravmann and Prussin, show both the checkered patterns including those crossed with the names of Allah or pseudo-writing in the same mode, as well as the pentacle and other *khawatim*. Their overall arrangements are highly similar to many of the *drapo*. A series of photographs by Herskovits taken near Mirabalais show a *houngan* making a funeral *vèvè* that includes the seventh of the *khawatim* signs.[49] Such design arrangements appear along with certain Hebrew phrases in the *grimoire* and books of alchemy, later in the handbooks of Freemasonry. (The role of Freemasonry in the Haitian Revolution and subsequent history is well recognized.) These symbols, then, are deeply integrated with one another, are intimately bound into matters of life and death and with the destiny of independence. Where such devices were similar to the banners of the French army, they could be easily conflated. Sapping power from an enemy by using his own symbols is a common strategy.

Might it be said that the uprising slaves made mojo banners in the face of Napoleon? Dessalines had Catherine Flon make the first Haitian flag—the most important *drapo* in any Vodou *hounfo*—from the deliberately torn French banner. Dessalines presented it on the feast of his patron *lwa*, Allou-mandia. Was that a mojo action too?

The Long Conversations and the Light from Brittany

The barque of Agwe sails into port. Just as all the participants in the slave and raw materials trade coexisted on the trans-Atlantic ships, so too do they coexist on the *drapo*, the *vèvè*, the *hounfo* objects and the services for the Kreyol *lwa*, Agwe. The modern *drapo* for the *lwa* typically feature a ship, often a sailing vessel, on whose side is written "Imamou." This Islamic word, written in Roman letters, signals the presence of a spiritual leader. It remains to be discovered when the word came to be written in a "foreign" orthography, or even when the first such *drapo* might have been fashioned. It is possible the writing came to Haiti with Muslims arriving in the late nineteenth century from the British West Indies. The *vèvè* probably predate the *drapo*, and some of them show the checkered pattern in addition to or instead of the word. The very earliest sign of the *lwa* may have appeared in Vodou *hounfo*; a small model ship suspended from the roof in a position near the *poto mitan*, the central pillar around which the sacred signs are written, ritual objects placed, and the ceremony circulates.

The ship would have been a highly charged symbolic object especially for those who had endured the Middle Passage. It was charged intensely for Breton seafarers, or for any who make their way in life over the waters. In Breton coastland chapels there is nearly always a model sailing vessel suspended from the ceiling or the rood-beam in close proximity to the chancel and altar. The antiquity of the practice is evident in the styles of the miniatures. Portuguese chapels often have such model ships, placed in the same way. Such votive ships may well have appeared in the numerous Christian chapels founded along the western African coasts. It is likely that mission churches in Cap Français (now Cap Haïtien), Port-au-Prince, St.-Marc, and other port cities could have included such votives.[50] An old *vèvè* for Ibo Lele, a *lwa* from Africa's Ibo people, used for funerals, depicts a ship as well, echoing this mortal significance. The Ibo were famous in colonial times for committing suicide rather than submitting to slavedom.

Not only would the ship be highly charged, but the stories and legends told would be as well, thus bringing up the matter of Agwe's consort, the Ezuli who takes the persona of Lasirenn. She is the mermaid, a figure that appears in African sea lore only after European arrival. In Brittany the famous mermaid is Arhes, the princess who disobeyed her father, the King of Is, by giving the royal keys to her diabolical lover, causing the entire city to sink into the Atlantic. Along this seismically active littoral, there is indeed a city that subsided, probably in Roman times, as there is a Roman road leading out into the ocean toward the site. Arhes appears in stone on the cornice of the parish church at Sizun, a look-alike of Lasirenn. It is Lasirenn who leads the faithful dead down under the sea to the mystical Lan Giné, sacred Guinea, the holy African homeland. ("Lan" is usually translated as a variant of "nan," Kreyol for "in" or "within." In Breton, "lan" means "holy" or "sacred," and the possibility that in this particular phrase it carries that meaning should be considered.)

Images of the ships, their prow ornaments, and their chains, the images and stories of the sea and of mysterious postmortem cities waiting to receive the faithful dead into the palaces of Is or Lan' Giné; the Long Conversation of Breton sailors telling of

Arhes the mermaid who sinks an entire city into the waters joins the actualities of awe and dread on the high seas; joins with the hope of return to the African homelands, down under the sea where Lasirenn will lead the dead to Agwe's holy kingdom. Pagan Arhes smiles down upon the Catholic faithful entering their church in Sizun. Agwe's sacred boat, "Imamou" and Lasirenn, or Ibo Lele, picked out in sequins, glitter sparks of light from Brittany, Mali and Ibo-land into a Vodou ceremony.

There are other snatches from the Breton Long Conversations and Theater; the heart, for example. The heart may be the oldest of all poetic synecdoches, standing easily for all of life, for the excitement of love, for the soul and spirit of a person. The "heart" shape, a fairly late graphic symbol bearing only slight resemblance to its anatomical referent, is the sign of the preeminent female *lwa*, Ezuli (or Erzulie). Not a symbol among Africans at the time of the Slave Trade, it probably came to Haiti in the duffels of sailors and the trunks of missionaries.

By the time of the Trade, images of the Sacred Heart of Jesus and the Immaculate Heart of Mary were to be found all over Europe, in churches and homes, and people carried them about as souvenirs. St. Marguerite-Mary Alocoque saw visions of the Sacred Heart of Jesus from 1673 to 1675, and a cult of the image was promoted by the Jesuits, eventually with papal approval in 1765. This Counter Reformation image elaborated upon hearts already current in Europe, including in books of alchemy. Anne Sauvy in *Le miroir du coeur* calls attention to its use in fifteenth-century images of Venus, like Ezuli, associated with love, and in playing cards, used for divination as well as gaming. Authorities forbade the use of anatomically correct hearts as too explicitly gory.[51] Contemporaneous example of the sacred hearts of Jesus or Mary range from a simple "heart" shape to depictions of its piercing by up to seven swords or punctured with a wreath of thorns, sometimes flaming, sometimes bleeding.

In Brittany the image of the heart became one of the most widely used of all images, especially in popular set of visual aids to the instruction of the faithful. Unusually, the laity themselves received instruction in using the posters to spread the instruction to their neighbors. Sauvy's book documents and explains how Michel LeNobletz devised the set of "moralised" heart images and promulgated them throughout Brittany. Remarkably durable, the images, as Sauvy shows, are in use in modernized versions in mission stations around the world as late as mid-twentieth century. The heart is sometimes shown surmounted by the face of a beautiful woman. Sometimes it includes implements of the Passion of Christ, and sometimes the animal emblems of the seven deadly sins. While none of the images shows the quadrille that frequently appears on the heart for Ezuli in Haiti, many are banded horizontally, indicating the Christian's progress to higher and higher states of goodness and moral virtue.

The images would have come into Haiti via the Breton sailors, instructed in their parishes by means of LeNobletz's images. It would have come in with Breton priests and religious as part of their teaching materials, or remade from memory once in the mission. It may have come quite early in the case of the sailors, later in the case of mission schools. Upon independence, Haiti was pariah in the Catholic Church, yet there were missions there, many of them Breton. Formal missions from the Brothers of Christian Instruction from the Grand Séminaire d'Haiti in Lampaul-Guimiliau in

Finistère came later. The Concordat was signed with the Papal See in 1860, and the second archbishop of Port-au-Prince (effectively of all Haiti) appointed shortly after, was the renowned Breton priest, Alexis-Jean-Marie Guilloux. Trained in Brittany, he was a great admirer of the work of Michel LeNobletz.[52] Considering the archbishop's concern for instruction and the founding of schools, it would be surprising not to find the LeNobletz images in all the Breton missions of Haiti by the late nineteenth century and into the twentieth.

Evidence that this is the case appears not only in the banners of Ezuli, but perhaps more indicatively in the paintings of the artist Préfète Duffaut who taught in his early youth at the Jacmel mission of the Breton Brothers of Christian Instruction. In an interview in 1996,[53] Duffaut was evasive about the source of his images, claiming, as he had to others, that he dreamed them or had visions of them and then he painted. What is the source for the dream images? Duffaut saw pictures at school. His paintings show precise elements that are features of the Breton moralized hearts. In one work from the 1950s, a woman who is probably at once the Virgin Mary and Ezuli stands upon a heart. Attached to it are instruments of Vodou and of the Passion of Christ. In the background of another painting is a Breton church of the sort not built in Jacmel, but appearing often in the moralized heart posters. Parallels abound, including rays streaming from the hands of Ezuli/Mary like those seen on images of Our Lady of Carmel or Our Lady of the Miraculous Medal. Several of Duffaut's later works show a similar woman, but this time drawing into her hands all the strands of a great web. She is Anansi, that trickster from the Akan people whose stories are in the Long Conversation of Diaspora people from Atlanta to Caracas.

The Breton priests especially may have brought their regional festival, the Pardons, to Haiti. Most parishes in Haiti have processions at special times of the year, but the Pardons have distinctive features that may have been taken up into Vodou. A Pardon service will include a salute of the banners to the four points of the compass, perhaps a vestige of the Long Conversation with the solstice and equinox services of the Druids. This is followed by the "kissing" of the banners with the crucifix around the compass points of the plaza in front of the church. Later there is a procession into the fields along stations of penance and forgiveness, or similarly toward the sea if the parish church lies shoreside. The orientation, the kissing of the banners, and the procession into the environment strongly parallel certain features of particular types of Vodou services, as does the use of splendid banners integral to the ceremony. (In Brittany too embroidery was held to be a sacred art, and Breton religious taught Haitian schoolgirls how to do the stitches.) In former times, the Pardon procession route might feature an elaborate "carpet" made of flower petals worked into a design of symbolic significance. Like the *vèvè*, and like the sand mandala of the Buddhists, it was meant to be walked upon, thus obliterated, by the faithful. The custom seems to have disappeared in Brittany, but in Haiti it may still be seen. As late as 1991, in Aquin (one of the parishes especially designated by Archbishop Guilloux for educational mission services) for the feast of Corpus Christi the youth of the parish created a design of flower petals depicting the Sacred Heart of Jesus on floor in front of a side altar. Had it appeared in a Vodou *hounfo*, it would have been immediately taken for a *vèvè* for Ezuli Dantor.

Breton and Muslim Long Conversations Tied

The Imamou of Agwe is one of several bindings of Breton and Islamic images and those of even older religious sensibilities. The sacred tree provides an example. Pre-Christian Bretons were mostly Druids, according primary significance to solar events, water sources, and trees. The sensibility is well grounded in environmental realism. Likewise, many traditional African religions held this set of worldly circumstances and objects to be paramount. In each case, the older habit of attention lies under the newer ones brought in on the one hand by Christians, on the other by Muslims. (There is a degree of traceable Moorish influence in Brittany as well, but that is another story.) The tree is taken up into West African Muslim practice, a sacred tree being a prominent feature of many Islamicized villages, and of certain graveyards. Minarets there usually include tree branches projecting from their surfaces, and a tree branch will be planted with each departed person, practices apparently borrowed from religions predating Islam and continuing as an undergirding.[54] In Brittany certain trees were honored from the time of the Druids, and continue to have certain mystical aura today. In Haiti they cut off ends of branches to allow free range to spirits, and tell that certain trees are the resting place of the powerful African *lwa*, Danbahlla. The motif of a sacred tree, including the national tree, the royal palm, is one of the most frequent of all Haitian painting, going back to works done in the first years of the Haitian Republic up until and including the present. It is a complex Kreyol motif with multiple penumbras from Africa, multiples from Brittany. At the time of the Haitian Revolution, the Tree of Liberty was an emblem of the people of France, and there was a counter emblem in Brittany, where liberty was seen by some as liberty from *any* French government, Royal or Republican.[55]

In *Dieu dans le vaudou Haïtien*, Laënnec Hurbon discusses the "dialectic of life and death around the symbol of the tree." He cites the force of the tree as a ubiquitous symbol that conjoins the potency of creation and its destruction, of chaos and order—a prevalent motif in Haitian life, literature, and religion.[56] The tale is told (and argued) that General Pétion, along with Dessalines, one of the heroes of the Haitian Revolution, insisted that the flag of the new country would include on its blazon the tree of liberty, the royal palm, which was the device symbolizing also his patron *lwa*, Aizan. Aizan, a female *lwa* associated with crops and with markets, is depicted as an older woman. It is she whose palm branches are shredded to cover the new initiates to Vodou, the *hounsi kanzo*. The historical Aizan was a powerful dowager queen who secured a large territory for her sons. In Haiti she came to be *mare*-tied to St. Anne, the grandmother of Jesus Christ and patron of an independent Brittany. Old feasts of St. Anne include processions with green branches, an event depicted in Haitian paintings.

Herskovits photographed services of offerings to trees, services observed in the present. Many Haitian paintings that depict Vodou services will include a tree with signs at its base, and those representing healings will often show the sick person being aided at the base of such a tree. A postcard-sized painting by J. C. Auguste done sometime before 1984 to be sold in the museum shop of the Musée d'Art of the College St.-Pierre in Port-au-Prince may be a version of such a healing. A *hounsi* seated by a tree on the right raises her arms in supplication. The tree is devoid of any leaves and the ends of its branches are cut. At the far left is an *assotor*, the largest of the Vodou ceremonial drums. Between *assotor* and *hounsi* rides a magnificent steed, half woman, her hair streaming in the wind. She carries a sword in her right hand and flaming scales of justice in the other. Around her neck is a Christian cross. That is a strange detail, for otherwise this appears to be not a Christian figure at all, but al-Burak, she who carried the Prophet Muhammed to the highest heaven. In Haiti many luminous strands come together in mutual illumination.

The *drapo* with checkered patterns, the power points that so closely resemble Islamic *khawatim*, the figures of Christian saints that are at the same time figures of Kongo, Ibo, Yoruba, Mende, Akan deities and heroes are fascinating material evidence of the Long Conversations, and also of another complex human equation: the creativity of the maker of objects, theater, songs that comprise the Long Conversations. It is the makers, the artists, who select the materials, the images, the styles. Moreau de St.-Mery's often cited descriptions of the slaves dancing around their "King" and "Queen" with brightly colored scarves aflutter is a theater without precedence in the European world. These slave artists had limited materials. They lived under repression. In the face of deprivation they created a visibility for their most cherished traditions out of new materials to suit the new condition. From slave times forward, military banners from Napoleon, op art from Seventh Avenue, Islamic magic squares, and Breton seafarers magic they took and take whatever is useful to reshape for purposes that suit the moment and the present companions, living and dead.

Notes

A substantial portion of the paper is based upon field observations in Haiti, Brittany and western Africa; review of film and photographic documents; and interviews with artists, *houngans*, dancers, historians, curators, and gallery directors in Haiti from 1981 to 1997.

1. Boite *T201* Archives Nationales, Paris, France. (These archives are now in Archives d'Outre Mer, Aix-en Provence.)
2. See, e.g., Terry Alford, *Prince Among Slaves* (New York: Oxford University Press, 1977). Alford tells the story of Abd al-Rahman, son of a Fulbe ruler, sold into slavery in 1788, impressive to both fellow slaves and to owners.
3. M. Elfasi, ed., *General History of Africa* (London: Heineman, UNESCO, University of California Press, 1988), 2:112–115.
4. LaBelle Prussin, *Hatumere: Islamic Design in West Africa* (Berkeley and Los Angeles, CA and London: University of California Press, 1986).
5. Anonymous, multiple authors, "Colloque sur Religions," in *Presence Africaine* (Abidjan and Paris: Editions Présence Africaine, 1961), 62.
6. See Jean Fouchard, *Marrons du Syllabaire* (Port-au-Prince: Editions Henri Deschamps, 1988), 22. On page 20, he cites early observations of slaves making ink and finding scraps of paper or even palm leaves to write down phrases that were then venerated. The phrases were most likely from the Qur'an. They could also have included magic squares, hatumere and khawatim.
7. Thomas Madiou, *Histoire d' Haiti Tome I; 1492–1799* (1847; repr., Port-au-Prince: Editions Henri Deschamps, 1989). See especially I:35.

8. Serge Fuertes, *Wol Vodou nan Bwa Kaiman* (Ottawa: Kawonabo, 1992).
9. J. R. Emmanuel Francius Julien, *La Cérémonie du Bois Caiman* (Port-au-Prince: Editions l'Ordre Nouveau, 1991), see 58, "Témoignage de Dédée Magritt," a recollection by a witness to the ceremony.
10. See Jean Fouchard, "Les Negres Islamistes," in *Marrons du Syllabaire*; Jean Fouchard, "Les Origines Africaines de l'Homme Haïtien," in *La Nouvelle Revue du Monde Noir* (Port-au-Prince: L'Institut d'Etudes et de Recherches Africaines d'Haiti, 1986), vol. 1, no.1, 181–223; *La Merengue, danse nationale d'Haïti* (Port-au-Prince: Editions Henri Deschamps, 1988). On pages 78–80 he notes the importance of thousands of slaves brought in from the eastern coasts and the island of Madagascar, attributing the name of Haiti's national dance to their language. Fouchard does not mention their Islamicization in this passage, but the Arab-Islamic presence along that coast was early and durable.
11. Roger Bastide, trans. Helen Sebba, *The African Religions of Brazil* ([Baltimore and London:Johns Hopkins University Press, 1978], originally published as *Les Religions Afro-Brésiliennes: Contributions à une sociologie des interpénétrations de civilisations* [Paris: Presses Universitaires de France, 1960]). See especially chapter 7, "Black Islam in Brazil."
12. David Lowenthal, *West Indian Societies* (London: Oxford University Press, 1972), 145. Cited in Abdullah Hakim Quick, *Deeper Roots: Muslims in the Caribbean Before Columbus to the Present* (Nassau, Bahamas and London: Association of Islamic Communities in the Caribbean and Latin America [AICCLA], 1990), 33.
13. The information concerning dietary prohibitions and the twice-daily ritual prayers comes from Donald Cosentino, in a conversation corroborating ideas that the Muslim influence might be more extensive in Haiti than had been supposed. Personal communication, October 1994. For further corroboration of this presence see below Dr. Rénald Clérismé, untitled, unpublished video documentation of Mandingue ceremony for deceased, taken near Cap Haitian, 1997, videocassette.
14. Quick, *Deeper Roots*, 21–22.
15. Ibid.
16. John Thornton, *Africa and Africans in the Making of the Atlantic World 1400–1680* (Cambridge: Cambridge University Press, 1992), 25.
17. Noel Q. King, *African Cosmos: An Introduction to Religion in Africa* (Belmont, CA: Wadsworth Publishing Company, 1986), 90.
18. William B. Cohen, *The French Encounters with Africans: White Response to Blacks 1530–1880* (Bloomington and London: Indiana University Press, 1980), 42–47.
19. There is inspiring, even touching, evidence from all over the Americas, including St. Domingue of Islamic captives reciting surahs by heart even years after captivity, and in face of severest repressions and strongest exhortations to convert to Christianity. See Fouchard, *Marrons du Syllabaire*, and Alford, *Prince Among Slaves*, for several examples. A German television broadcast in 1990 showed Qur'an boards in a "Museum of Slavery" then open in Charleston, South Carolina. Attempts to examine the boards met first with the closure of the museum, then of their apparent sequestering in a private home off the coast. Making these invaluable documents accessible again is highly desirable.
20. Pierre Pluchon, *Vaudou sorciers empoisonneurs de Saint-Domingue à Haïti* (Paris: Editions Karthala, 1987), 14–15.
21. René Bravmann, *African Islam* (Washington, DC and London: Smithsonian Institution Press and Ethnographica, 1983). See illustration no. 40, 56.
22. Clérismé, untitled unpublished video documentation of Mandingue Ceremony for deceased in 1997, shown privately during the May, 1998 Congress of Santa Barbara, *The Spirit and the Reality: Vodou and Haiti*, at Brooklyn College. I am grateful to Dr. Clérismé who very graciously offered to show me this tape upon hearing my paper in which I

suggested that phrases from the Qur'an, prayers, and praise songs might be embedded in the *langaj* of Vodou. The participants in the ceremony clearly were chanting "Allah al-illah." They wore white garments which may, like certain candomble costumes in Brazil, be related to Muslim burial costume.

23. Daniel Cozort, *The Sand Mandala of Vajrabhairava* (Ithaca, NY: Snow Lion Publications, 1995). Centerfold, colored illustrations of the mandala and the three-dimensional computer generation.
24. Karen McCarthy Brown, *The "Veve" of Haitian Vodou: A Structural Analysis of Visual Imagery*, thesis submitted to Temple University Graduate Board, 1975. (Ann Arbor: University Microfilms, 1976). See especially chapter 4, "The Center and the Edges," in which she discusses the relationship of the symbols to the arrangements in Haitian society.
25. LeGrace Benson, "Observations on Islamic Motifs in Haitian Religious Art," *Journal of Caribbean Studies*, (1 and 2) (Winter/Spring 1993). In this article, the squares are called "magic squares." Further research has revealed them to be "destiny squares" more specifically.
26. Anna A. Wexler, "The Artistry of a Haitian Vodou Flagmaker," in *Sacred Possessions: Vodou, Santeria, Obeah and the Caribbean*, ed. Marguerite Fernández Lomos and Lizabeth Paravasini-Gebert (New Brunswick: Rutgers University Press, 1997), 69.
27. Delores Yonkers, *Sequinned Surfaces: Vodoun Flags from Haiti.* Exhibition Catalogue (Northridge: California State University, 1991). Unpaginated catalogue for exhibition curated by Yonkers.
28. Ibid.
29. Patrick Polk, "Sacred Banners and the Divine Cavalry Charge," in *Sacred Arts of Haitian Vodou*, ed. Donald Cosentino (Los Angeles: UCLA Fowler Museum of Cultural History, 1995), 325–347.
30. Constance Abikuba, personal conversation concerning her father and other men in the family. She told that they thought the game was "very old," and that they said the Islamic people to the north had brought it in.
31. Peter Adler and Nicholas Barnard, *Asafo! African Flags of the Fante*. Exhibition catalogue, heavily illustrated in color (New York: Thames and Hudson, 1992). Flags viewed at Johnson Museum of Art, Ithaca, NY, February, 1997. Variations on the checkered pattern and variants as well as depictions of a chess board or game in progress appear on a large number of the flags.
32. Bravmann, *African Islam*, 61. See illustration 44.
33. Erika Bourguignon, unpublished photographic documentation, 1997.
34. J. E. Cirlot, *A Dictionary of Symbols* (New York: Philosophical Library, 1962), 43.
35. Henry Gough and James Parker, *A Glossary of Terms Used in Heraldry* (1894; repr., Oxford and London: Gale Research, 1996), 103.
36. Maude Southwell Wahlman, "African Charm Traditions Remembered in the Arts of the Americas," in *African Impact on the Material Culture of the Americas*, conference proceedings (Winston-Salem: Old Salem, 1998). Figure 88, 9.
37. Eli Leon, "African Influence on the American Block-Style Quilt," in *African Impact on the Material Culture of the Americas*, conference proceedings (Winston-Salem: Old Salem, 1998).
38. Susan Tselos, "Haitian Drapo Vodou; Imagery, Ritual and Perception," in *Sacred and Ceremonial Textiles* (Chicago: Textile Society of America, 1995), 58–67.
39. Drouin de Bercy, *De St. Domingue, de ses guerres, de ses ressources* (Paris: Hoquet, 1814), cited by Emmanuel C. Paul, *Panorama du Folklore Haïtien* (Port-au-Prince: Imprimerie de l'Etat, 1962), 143.

40. Several owners of boutiques in Haiti confirm this. A private video tape done in the 1960s documents a discussion between a *houngan*/artist and director of an atelier, and other shop managers relate similar events. A musical group of the earlier 1960s commissioned banners for display behind their group, requesting that they have authentic Vodou symbols. Anna Wexler, in an unpublished doctoral dissertation, Harvard University, 1998, describes such a process from the standpoint of *houngan*/artist Clotaire Bazile. It is clear from reports and especially from Wexler's investigations that traditions are carefully continued, despite creative use of available new materials and new graphics styles.
41. See, e.g., illustration 32 in Harold Courlander's *The Drum and the Hoe: Life and Lore of the Haitian People* (1960; repr., Berkeley, LA and London: University of California Press, 1973), where the *vèvè* drawn on the occasion of a bull sacrifice for Simbi shows the checkered pattern on the flanks of the bull. The design is highly similar to that used by the Islamicized Senufo for the checkered bush-cow costume, woven of raffia and painted with checks or *fusilly*. Illustrations in Déita [Mercedes Foucard Guignard], Objets au Quotidien: Art et Culture Populaire en Haïti Port-au-Prince: Imprimerie Le Natal S. A., 1993; Chapter 16 concerning amulets and talismans, 103–112 show the pattern on Boussou Trois Cornes and one of the Agwe boats.
42. Milo Rigaud, *Secrets of Voodoo*, trans. Robert B. Cross (1953; repr., San Francisco: City Lights Books, 1985), 148.
43. The unpublished photographs are currently held in the African-American Museum section of the Smithsonian Institutions, property of the Estate of Melville Herskovits. Scholars may review the photographs upon request. Dr. Maude Wahllman, a former student of Dr. Herskovits kindly allowed me to review and take notes from a notebook of photocopies of the documents, June, 1998.
44. Bravmann, *African Islam*. See illustration no. 37, 54.
45. Clotaire Bazile, interview, Brookline, MA, September 1993.
46. Rigaud, *Secrets of Voodoo*, 151.
47. Clérismé, untitled, unpublished video documentation of Mandingue ceremony for deceased shown in 1997.
48. Bravmann, *African Islam*, 50.
49. Herskovits photographic archives, Smithsonian Institutes, African-American Museum, pictures 45, 46, 47.
50. A search through numerous inventories of churches and the personal belongings of missionaries includes only items that had some monetary value, and lack mention of a range of objects that probably were there, but not part of the formal valuation. Votive objects of all sorts were common in the era of the slave trade and into the present, but seldom appear in inventories. Exceptionally, a votive in precious metals and stones might make the list. Other evidence should be sought.
51. Anne Sauvy, *Le miroir du coeur* (Paris: Les Editions du Cerf, 1989), 48–50. The entire work is an invaluable resource for studying the imagery of the heart and missions instructions.
52. P. A. Cabon, *Mgr. Alexis-Jean-Marie Guilloux, Deuxième Archevêque de Port-au-Prince* (Lampaul-Guimiliau and Port-au-Prince: Archevêché de Port-au-Prince, 1929), 23. Cabon indicates Guilloux took LeNobletz as one of his role models.
53. Duffaut, interview, Centre d' Art, Port-au-Prince, Haiti, November 1996. The artist brought in current pieces that had none of the imagery under discussion here, but rather were versions of his mountain villages. Some of his other village or city images contrast the city of good with the city of evil, and do show elements of the good/evil moralizations of the Breton teaching materials.
54. Prussin, *Hatumere*, 168.

55. A regional museum in Plougastel, Brittany in 1989 had a special exhibition for the Bicentennaire of the French Revolution. Among other aspects, there were several documents and prints showing the multifariousness of the liberty tree in Brittany. It was clearly an important emblem, contested by pro-French and pro-regionalists.
56. Laënnec Hurbon, *Dieu dans le vaudou Haïtien* (Port-au-Prince: Editions Henri Deschamps, 1987), 129–137.

Bibliography

Adler, Peter, and Nicholas Barnard. *Asafo! African Flags of the Fante*. Exhibition catalogue, heavily illustrated in color. New York: Thames and Hudson, 1992.

Alford, Terry. *Prince Among Slaves*. New York: Oxford University Press, 1977.

Archives d'Outre Mer. Archives Nationales de France, Aix-en-Provence, France. These archives include all Moreau de St. Méry's writings, and collected materials for St.-Domingue as well as the other French Caribbean colonies. There are documents of notaries; inventories; military records; some family records and various official correspondences. Materials directly referring to religious practices of any kind are scant. The Archives include documents from other nationalities having contact with the French colonies, including some interesting and informative engravings of flora, fauna, plantations, sugar planting, and processing.

Bastide, Roger. *African Religions of Brazil*. Trans. Helen Sebba. Baltimore and London: Johns Hopkins University Press, 1978. Originally published as *Les Religions Afro-Brésiliennes: Contributions à une sociologie des interpénétrations de civilisations*. Paris: Presses Universitaires de France, 1960.

Benson, LeGrace. Observations on Islamic Motifs in Haitian Religious Art. *Journal of Caribbean Studies*, 9 (Winter/Spring 1993): 59–66.

———. Some Breton and Muslim Antecedents of Vodou Drapo. In *Sacred and Ceremonial Textiles*. Proceedings of the Fifth Biennial Symposium of The Textile Society of America, 68–75. Chicago: Textile Society of America, 1996.

de Bercy, Drouin. *De St. Domingue, de ses guerres, de ses ressources*. Paris: Hoquet, 1814. Cited in Emmanuel C. Paul, A Processus de la formation du Vodou. In *Panorama du folklore Haitien*, 238–239. Port-au-Prince: Imprimerie de l'Etat, 1962.

Blakely, Thomas D., Walter E.A. van Beek, and Dennis L. Thompson. *Religion in Africa*. Monograph Series of the David M. Kennedy Center for International Studies at Brigham Young University. London: James Curry and Portsmouth, NH: Heinemann, 1994.

Bravmann, René A. *African Islam*. Washington, DC and London: Smithsonian Institution Press and Ethnographica, 1983.

———. *Islam and Tribal Art in West Africa*. 1974. Repr., Cambridge: Cambridge University Press, 1994.

Bourguignon, Erika. Unpublished photographic documentation. Haiti, 1944–1954.

Brown, Karen McCarthy. *The Vèvè of Haitian Vodou: A Structural Analysis of Visual Imagery*. Ann Arbor: University Microfilms, 1976.

Cabon, P. A. *Mgr. Alexis-Jean-Marie Guilloux, Deuxième Archevêché de Port-au-Prince*. Lampaul-Guimiliau and Port-au-Prince: Archevêque de Port-au-Prince, 1929.

Centre d' Art. Uncatalogued archives include letters, photographs, and newspapers. Collection of works from 1947 and current exhibitions. Port-au-Prince, Haiti.

Cirlot, J. E. *A Dictionary of Symbols*. New York: Philosophical Library, 1962.

Clérismé, Rénald. Untitled, unpublished video documentation of Mandingue ceremony for deceased, taken near Cap Haitian, 1997, videocassette.

Cohen, William. *The French Encounters with Africans: White Response to Blacks, 1530–1880.* Bloomington and London: Indiana University Press, 1980.
Colloque sur les religions. Conference proceedings. Abidjan and Paris: Les Editions Présence Africaine, 1961.
Cosentino, Donald J., ed. *Sacred Arts of Haitian Vaudou.* Catalogue of exhibition circulated to museums in the United States. Heavily illustrated. Los Angeles: Fowler Museum of the University of California at Los Angeles, 1995.
Courlander, Harold. *The Drum and the Hoe: Life and Lore of the Haitian People.* 1960. Repr., Berkeley, LA and London: University of California Press, 1973.
Cozort, Daniel. *The Sand Mandala of Vajrabhairava.* Ithaca, NY: Snow Lion Publications, 1995.
Déita, Mercédès Guinard. *Objets au quotidien; Art et culture populaire en Haïti.* Port-au-Prince: l'Imprimerie LeNatal S. A., 1993.
———. *La Légende des Loa du Vodou Haïtien.* Port-au-Prince: Private Printing, 1993.
Delbeau, Jean-Claude. *Société, culture et médicine populaire traditionnelle.* Port-au-Prince: Imprimerie Henri Deschamps, 1990.
Deren, Maya. *Divine Horsemen.* London and New York: Thames and Hudson, 1953.
Desquiron, Lilas. *Racines du Vodou.* Port-au-Prince: Editions Henri Deschamps, 1990.
Dunham, Katherine. *Island Possessed.* Chicago and London: University of Chicago Press, 1969.
Elfasi, M., ed. *General History of Africa.* Vol. 2. London: Heineman, UNESCO and University of California Press, 1988.
Fombrun, Odette Roy. *Le Drapeau & les armes de la République d'Haiti.* Port-au-Prince: Editions Deschamps, 1987.
Fouchard, Jean. *Marrons du Syllabaire.* Port-au-Prince: Editions Henri Deschamps, 1988.
———. *La Merengue, danse nationale d'Haiti.* Port-au-Prince: Editions Henri Deschamps, 1988.
———. Les Origines Africaines de l'homme Haïtien. In *Nouvelle Revue du Monde Noir.* Port-au-Prince: L'Institut d'Etudes and de Recherches Africaines d'Haiti, 1986.
Fuertes, Serge. *Wol Vodou nan Bwa Kaiman.* Ottawa: Kawonabo, 1992.
Glaze, Anita J. *Art and Death in a Senufo Village.* Bloomington: Indiana University Press, 1981.
Gough, Henry, and James Parker. *A Glossary of Terms Used in Heraldry.* 1894. Repr., Oxford and London: Gale Research, 1996.
Herskovits, Melville. *Life in a Haitian Valley.* New York and London: Alfred A. Knopf, 1937.
———. African American Museum Photographic Archives. Smithsonian Institutes: Washington, DC.
Holas, B. *Arts Traditionnels de la Côte d'Ivoire.* Tours: Imprimerie Rellure, 1967.
———. *L'Art Sacre Sénufo.* Abidjan: Les Nouvelles Editions Africaines, 1985.
Hurbon, Laënnec. *Dieu dans le Vodou Haïtien.* Port-au-Prince: Editions Henri Deschamps, 1987.
———. *Le Barbare imaginaire.* Port-au-Prince: Editions Henri Deschamps, 1987.
Journal of Caribbean Studies. Lexington, KY: Association of Caribbean Studies.
Journal of Islamic Studies. Oxford: Oxford University Press.
Journal of Haitian Studies. Boston: Haitian Studies Association.
Julien, J. B. Emmanuel Francius. *La Cérémonie du Bois Caiman.* Port-au-Prince: Editions l'Ordre Nouveau, 1991.
King, Noel Q. *African Cosmos: An Introduction to Religions in Africa.* Belmont, CA: Wadsworth Publishing Company, 1986.
Kroch Rare Book Library, Cornell University Libraries, Ithaca, NY. The collections include, inter alia those devoted to Freemasonry in Europe and the Americas, the French Revolution, Slavery, Witchcraft, and chess. The Witchcraft collection has multiple copies

of the numerous *grimoires* published in Europe from the fifteenth through the nineteenth centuries, examples of which still appear in contemporary Haiti.

Leon, Eli. African Influence on the American Block-Style Quilt. In *African Impact on the Material Culture of the Americas*. Conference Proceedings. Winston-Salem: Old Salem, 1998.

Lerebours, Michel-Philippe. *Haïti et ses peintres de 1804 à 1980*. Vols. 1 and 2. Port-au-Prince: Imprimeur II, 1989.

Lowenthal, David. *West Indian Societies*. London: Oxford University Press, 1972.

Madiou, Thomas. *Histoire d'Haiti Tome I; 1492–1799*. 1847. Repr., Port-au-Prince: Editions Henri Deschamps, 1989.

Maximilien, Louis. *Le Vodou Haitien; rites radas-canzo*. Port-au-Prince: Imprimerie H. Deschamps, 1945.

Métraux, Alfred. *Le Vaudou Haïtien*. 1958. Repr., Paris: Editions Gallimard, 1968.

Musée d'Art Haïtien. Collections from 1947 and exhibitions of current works. Port-au-Prince.

MUPANAH. The national museum includes a permanent historical exhibition as well as art works from all periods and current exhibitions. Port-au-Prince.

Nouvelle Revue du Monde Noir. Port-au-Prince: L'Institut d'Etudes et de Recherches Africaines d'Haiti (IERAH).

Olmos, Fernandez, and Lizabeth Paravisini-Gebert, eds. *Sacred Possessions; Vodou, Santería, Obeah and the Caribbean*. New Brunswick: Rutgers University Press, 1997.

Paul, Emmanuel C. *Panorama du Folklore Haïtien*. Port-au-Prince: Imprimerie de l'Etat, 1962.

Pluchon, Pierre. *Vaudou sociers empoisonneurs de Saint-Domingue à Haïti*. Paris: Editions Karthala, 1987.

Polk, Patrick. Sacred Banners and the Divine Cavalry Charge. In *Sacred Arts of Haitian Vodou*, ed. Donald Cosentino, 325–347. Los Angeles: UCLA Fowler Museum of Cultural History, 1995.

Prussin, LaBelle. *Hatumere: Islamic Design in West Africa*. Berkeley and Los Angeles, CA and London: University of California Press, 1986.

Quick, Abdullah Hakim. *Deeper Roots: Muslims in the Caribbean before Columbus to the Present*. Nassau, Bahamas and London: Association of Islamic Communities in the Caribbean and Latin America (AICCLA), 1990.

Qur'an. Arabic Text, English rendering Abdul Aziz Kamal. Lahore, Pakistan: Islamic Publications, 1982.

Quattara, Abdoulaye. *La Danse* [de Côte d'Ivoire]. Abidjan: Les Nouvelles Editions Africaines, 1985.

Rigaud, Milo. *Secrets of Voodoo*. Trans. Robert B. Cross. 1953. Repr., San Francisco: City Lights Books, 1985.

Sauvy, Anne. *Le miroir du coeur: quatre siècles d'images savantes et populaires*. Paris: Les Editions du Cerf, 1989.

Searing, James F. *West African Slavery and Atlantic Commerce: The Senegal River Valley, 1700–1860*. African Studies Series No. 77. Cambridge: Cambridge University Press, 1993.

Thornton, John. *Africa and Africans in the Making of the Atlantic World, 1400–1680*. Cambridge: Cambridge University Press, 1992.

———. *The Kingdom of the Kongo: Civil War and Transition, 1641–1718*. Madison: University of Wisconsin Press, 1983.

Tselos, Susan. Haitian Drapo Vodou: Imagery, Ritual and Perception. In *Sacred and Ceremonial Textiles*. Chicago: Textile Society of America, 1995.

Wahlman, Maude Southwell. African Charm Traditions Remembered in the Arts of the Americas. In *African Impact on the Material Culture of the Americas*. Conference proceedings. Winston-Salem: Old Winston, 1998.

Wexler, Anna A. I am going to see where my Oungan is: The Artistry of a Haitian Vodou Flagmaker. In *Sacred Possessions: Vodou, Santería, Obeah and the Caribbean*, eds. Marguerite Fernández Olmos and Lizabeth Paravisini-Gebert. New Brunswick: Rutgers University Press, 1997.

———. For the Flower of Ginen: The Artistry of Clotaire Bazile, A Haitian Vodou Flagmaker. PhD dissertation, Harvard University, 1998.

Yonkers, Dolores. *Sequined Surfaces: Vodoun Flags from Haiti*. Exhibition Catalogue. Northridge: California State University, 1991.

Chapter 11

Vodou in Haitian Literature

Carrol F. Coates

Senator Jesse Helms, that staunch defender of "American values" and Chair of the Senate Foreign Relations Committee, recently accused the United States Agency for International Development (USAID) of "subsidizing witchcraft," according to a report by Don Bohning in the *Miami Herald* (March 15, 1999). It is apparent in Bohning's report that the honorable senator equates Vodou with "witchcraft." From his own standpoint, Helms was undoubtedly justified in that the 1995 "Annual Report" of the International Planned Parenthood Federation in Haiti stated that it had undertaken an educational "campaign to reach voodoo followers with sexual and reproductive health information." I admit that I struggle with my demon to keep from throwing hard, sharp objects at the television set when the senator appears on news reports, but his equation of Vodou with sorcery is probably not far from the view held by a significant number of U.S. citizens, especially those who are capable of coining a variety of satirical and deprecatory terms using "voodoo" as an adjective implying irrational and manipulatory procedures—giving us such useful neologisms as "voodoo economics."

Unlike Karen McCarthy Brown, whose *Mama Lola* is listed in my "Bibliography on Vodou," I am not an initiate, but I feel that I am close to her position in that I too operate in a space between that of the faithful of Vodou, those who "follow the spirits," and that of a white North American, raised and educated in the Oklahoma plains. I have come closer to appreciating Vodou as a legitimate religion over a period of some ten years as I undertook translating Haitian literature from French to English and, eventually, teaching courses on Haitian Literature and Culture, as I have done several times at Binghamton University. I discovered, gradually, that many, if not all Haitians, have had some kind of contact with Vodou, at least through other family members. A majority of Haitian novelists make reference to Vodou, and I will give a rapid analysis of several works in which that culture is either central or at least an important aspect of the work. I hope to articulate a suggestion or two for further reflection as to what functions various writers may be assigning to Vodou in their creative work.

It is necessary to take a distance, first of all, from the prevalent Anglo-Euro-American view that Vodou is simply witchcraft, sorcery, or "black magic." The tenth edition of *Merriam Webster's Collegiate Dictionary* still retains the pejorative association with "necromancy," hexes, and sorcery. "Voodooism" is defined as "the practice of witchcraft." I might note, without further comment, that the terms "voodoo" and "voodooism" fall between the words "vomitory" and "voracious"—as if Senator Helms's genie had some mystical control over the "accidents" of *Webster*'s selection and alphabetization of its lexicon.

Even *Webster's* brief descriptive phrase is less objective than it might appear: "a religion that is derived from African polytheism and ancestor worship and is practiced chiefly in Haiti." Traditional African and Caribbean religions are not "polytheistic" in the Western sense and a number of other prejudicial terms are still commonly used—"animism," "fetishism," "paganism," "primitivism," among others. Filmmakers and popular novelists continue to conjure up images of cannibalism and zombification as prevalent practices.

Let us turn to what Vodou is and distinguish it from the simple practice of sorcery and witchcraft.

In her recent study, *Aspects éducatifs et moraux du Vodou haïtien*, Claudine Michel (UC Santa Barbara) characterizes Vodou as a "global vision of the world" (1995, 24). She states that Vodou is "a way of life for the majority of people in Haiti" (1995, 28). Karen McCarthy Brown sees Vodou as "the system [Haitians] have derived to deal with the suffering that is life" (1991: 10). From these, and many other statements by credible researchers and observers, we might derive two aspects of Vodou: it emerged, early on in Haiti, as a traditional religion, preserved, modified, and systematically disguised in order to survive under the daily cruelty and deprivation of the plantation system. Vodou was at once a view of the world that evolved with many different local adaptations in accordance with the experience of the communities of slaves and it served (1) to preserve memories and traditions of the Africa from which the slaves or their ancestors came; (2) to develop the capacity to endure the hardships of daily existence; and (3) eventually, to rally fellow slaves to revolt against the inhumanity of the plantation system.

In *Haiti: The Breached Citadel*, Patrick Bellegarde-Smith states that, in Vodou,

> everything in the world—be it plant, animal, or mineral—shares basically similar chemical, physical, and/or genetic properties. This unity of all things translates into an overarching belief in the sanctity of life, not so much for the *thing* as for the *spirit* of the thing. The cosmological unity of Vodun further translates into a vaunted African humanism in which social institutions are elaborated and in which the living, the dead, and the unborn play equally significant roles in an unbroken historical chain. (1990, 12)

This is important in that it underscores the basic trait of Vodou as a "cosmology," a comprehensive view of the world that involves a basic respect for all being, whether animate or inanimate, human or animal, plant or mineral. Bellegarde-Smith further stresses that sex and gender are not prejudicial matters for those who serve the spirits: both men and women can be priests; homophobic prejudices and violence are absent (13).

Vodou has no Bible or Qur'an—that is, no written compilation of sacred writings. It was a simple, perhaps natural matter for the slaves to accept the images of Catholic saints as equivalent to the *lwa* or spirits of Vodou and the chromolithographs of Christian saints are often found on Vodou altars. As numerous people, both scholars and those who serve the *lwa*, have pointed out, this does not mean that the servants of the spirits equate the saints with the *lwa*. The spirits, as well as symbols and the actual spaces of the *ounfò*, or Vodou places of worship, may change and new spirits may appear and be welcomed on a local basis, so that there is no comprehensive compendium of *lwa*, for instance.

Let me simply suggest a rapid list of the realms of knowledge and belief that are involved for the faithful in Vodou. Vodou is a religion, a set of beliefs about the nature of being and the universe. Vodou involves complex and highly imaginative forms of art that have grown up through sincere religious practices—visual arts and sculpture, dance, music, and dramatic representation. Since the *ounfò* is fundamentally a family affair—Vodou has no pope and no bishops or archbishops—it involves social values and morality for the individual, the family, and the surrounding or conjoined community. Especially in the rural regions of Haiti, but also in the cities, doubtless, Vodou has its influence on commerce and business—thus, a national impact on the economy.

Two important functions of Vodou remain to be stressed. First of all, Vodou is "a system whose purpose is to minimize pain, avoid disaster, cushion loss, and strengthen survivors and survival instincts" (Brown, 1991). It is a religion of healing: the *manbo* and the *oungan* are basically healers who know and deal with herbs and specifics. There is a forthright and unambiguous belief in the power of the spirits and the need to deal with them generously, respectfully. Second, Vodou is a system for education, and this is the basic message of Claudine Michel's overview of the "educational and moral aspects of Haitian Vodou." Just as it involves respect for everything that exists in the universe, Voudou is an important system for communicating the beliefs of the family and the community to the young. The values taught include respect for the elders and for the ancestors, from which the distorted view of *Webster's* can derive the simplistic trait of "ancestor worship."

It is true that Haitians belonging to various fundamentalist sects have, along with Jesse Helms, demonized the faithful of Vodou as Satan-worshippers, sorcerers, and werewolves. There is a grain of truth to this in that some *oungan* and *manbo* have undoubtedly practiced with both the right and left hands—that is, operated as both healers and sorcerers. I occasionally consider contacting an *oungan* to see whether I cannot concoct a deal to put a hex on certain individuals whom I consider a disgrace to the human race, but I have resisted the temptation for fear of ending up in disfavor with the spirits, who can be merciless when offended.

One of Hollywood's favorite topics is that of zombi. Many people have reported meeting zombi and I have on rare occasion met a person who struck me as being zombified, as showing no sign of having a soul, no spark of life in the eyes. I can recommend one serious study for anybody interested who reads French. It is entitled *Le phénomène zombi* and was written by a Haitian neuropsychiatrist living in France, Dr. Yves Saint-Gérard, who recognizes, along with the sensational Wade Davis, that there are pharmaceutic substances that can reduce vital functions to a level that makes life undetectable. Other products might reduce the aggressivity of victims being held by an individual *bòkò, oungan*, or members of a secret society. Dr. Saint-Gérard ends up recognizing that the Haitian phenomenon of zombification is a combination of the possibility of drugs or pharmaceutical products, psychological intimidation or hypnosis, a belief in the spirits, a certain level of social acceptance of the reality of such practices, and, finally but perhaps most importantly, the political abuse of the integrity of human rights and the manipulation of widespread social fear and apprehension. In response to a series of postings on werewolves on the Haiti e-mail net, Leslie Desmangles posted the basic theme that I have taken in my introduction, namely that Vodou is basically a legitimate religion. He makes the distinction between the main traditions, which include service for the spirits and healing, and the "little traditions" such as zombification that are not part of Vodou properly speaking. The zombi phenomenon "is not part of Vodou and never was, as Max Beauvoir himself has often noted. If it exists, zombie-making should be regarded as a 'folk'. . . religious practice that derives from Vodou beliefs but is not part of them." Desmangles's message is basically a criticism of the book by Wade Davis, *The Serpent and the Rainbow*, which is faulted for viewing zombification as part of Vodou (Desmangles, e-mail, April 1, 1999).

Vodou and those who serve the spirits have been perpetual objects of hatred, fear, and persecution, from colonial times to the present. The practice of African religions was forbidden on the plantations and the colonialists lived in justified fear of poisoning and revolt by the slaves. The Haitian Revolution began with a Vodou ceremony at the Bwa Kayman, in the mountains close to the Dominican border and quickly spread throughout the colony. Following Haitian Independence, relations between the Haitian state and Rome were not formalized until the Concordat was signed in 1860, under President Geffrard. There were periodic efforts by the Roman clergy to counteract the importance of Vodou in Haiti, most importantly under President Élie Lescot in the anti-superstition campaign of 1941–1942. Many, if not all, Haitian presidents practiced Vodou, along with Christianity, whether openly or secretly. Jacques Stephen Alexis says that President Vincent went, during his presidency, to consult a famous *oungan*. Élie Lescot's grandmother took her newborn grandson to the *oungan*, Papa Pierre, who predicted that that little Élie had a political future (Rigaud 1953b, 48). It is well known that François Duvalier followed the spirits and that some of his close associates were *oungan*, but Karen McCarthy Brown says that he was himself an *oungan* (1991, 95). Vodou became such an instrument of repression and political abuse with the Duvaliers, father and son, that there was an eruption of vengeful attacks on *oungan, manbo*, and the Vodou sanctuaries, following the ousting of Baby Doc in 1986.

* * *

I am going to survey a selected number of works in which Vodou plays a significant role. The first group includes works in which sorcery and the abuse of Vodou by the head of state during the two Duvalier regimes. The second category includes just one poem where the vision of Vodou becomes a revolutionary discourse. In the third category, I include several novels in which Vodou is represented as an important aspect of the people's vision of the world. Finally, there are four novels in which Vodou or the supernatural constitute the most important informing reality.

A number of works have presented Vodou as a tool in the hands of the Duvaliers, father and son. A capital scene in René Depestre's *Festival of the Greasy Pole* is the ceremony that is organized by a *bòkò* in the National Palace in order to seek the support of the spirits in bringing about the defeat of the rebel Henri Postel in the contest of the greasy pole. During the nocturnal ceremony, the naked President Zoocrates (the caricature of Papa Doc) is carried around the hall of the Palace on the backs of sorcerer, Simon-Seven-Days, various *tonton makout*, and Monsignor Wolgondé (a caricature of the Archbishop, François-Wolf Ligondé). Needless to say, Depestre's hero will win the contest by reaching the top of the greased pole, only to be mortally wounded when he grabs a machine gun and attacks the President-for-Life. Gérard Étienne also portrayed Duvalier's close advisors as *tonton makout* and *oungan* at the same time in his fourth novel, *La Reine Soleil levée* (1988). The late Franck Fouché wrote a bitterly satirical play, *Général Baron-la-Croix*, conceived as a ceremony in which the impotent president (Duvalier, of course) concludes a bargain with Baron Lacroix to mount his wife in order to give him an heir and successor.

Franketienne's *Dezafi*, the first novel written entirely in Kreyòl, was rewritten in French as *Les affres d'un défi* (1979) (in the grip of a challenge). Franketienne explains the meaning of his original Kreyòl title in the glossary of *Les affres*:

> *Dézafi:* a sort of fair organized in certain Haitian provinces; along with orgiastic festivals, cockfights are the main attraction. The word "dezafi," in a broader sense, would mean a great gathering, a popular movement, the ferment of a crowd. (1979, 232)

The story that is told, in both the Kreyòl and the French versions, is that of a group of men who have been zombified by the *bòkò*, Saintil. They are made to walk through the country at night to work in various rice paddies. They are underfed and, for the least word or protest, are severely punished, by beatings, maiming, and even execution. The scene alternates between the inhuman existence of the zombies, disorderly crowd scenes at cockfights, and interludes in which we are witness to the anguish of several characters who are in hiding in Port-au-Prince. In fact, however, there is no traditional plot, with full character development and dramatic sequencing. The segmentation of the novelistic discourse is marked by three fonts—Roman, italics, and bold face, each representing a different "voice." This is an essential technique of Franketienne's "spiralism," in which he shifts from one voice (or set of voices) and one scene to another without transition or explanation.

It is fairly evident, in this 1979 French version (the original Kreyòl text was published in 1975), that the zombi, as well as the various characters who are living in mortal fear as they hide in their apartments, represent Haitians under the regime of Baby Doc. Saintil could be a figure for the young President-for-Life, not by direct portrayal, but by his brutal exercise of power and the mercilessness with which he commands his daughter and his henchman, Zofer, to control and punish the zombi. The striking feature of the text, however, is that it foregrounds the alternately coherent and incoherent murmur of voices of the zombified. The reader is drawn into this confused buzz of suffering and muted revolt that will be broken toward the end of the novel when the *bòkò*'s daughter, Sultana, falls in love with a stalwart zombi and puts salt into his stew, the antidote that releases the victim from the state of zombification.

Franketienne does not himself serve the spirits, according to his own testimony. He told me he regretted not having been initiated, however, and I read his presentation of sorcery, in *Les affres*, as strictly a symbol of religion and magic used as instruments of fear and repression by Baby Doc, or perhaps, more broadly, by both Duvaliers. I see no sign of an attack on Vodou as a religion.

Stanley Péan, a young Haitian writer living in Québec City, likes to establish a fictional world that partakes both of historical reality and of the supernatural. The scene of the first, introductory part of *Zombi Blues* (1996) is Port-au-Prince in 1971, just prior to the death of François Duvalier. At a moment when she was starving, a young Haitian mother agreed to let Mèt Minville have her *marasa*, twins, in exchange for food. The woman's dying husband told her to get the boys back. She is wounded, however, either by the *makout* or by the henchmen of Mèt Minville. Evading her pursuers, she gets a Canadian woman to take charge of the one baby that she has gotten away from Mèt Minville. The Haitian mother dies and the rest of the novel takes place in Montréal in 1986–1987.

There is not time to deal with the intricately woven plot of the novel. Let it suffice to say that two apparently unrelated groups of characters turn out to be associated with the two separated *marasa*, who are now young men. Gaby d'ArqueAngel is a jazz musician working in Montréal. Caliban, or "Gran Blan," is the albino servant of Mèt Minville, who has come to Montréal to seek asylum, after the *dechoukaj* of Baby Doc, in 1986. The reader eventually finds out that Gaby's visions and Caliban's headaches result from their *marasa* relationship and the fact that, when they were babies, Mèt Minville administered a potion that gave each virtually superhuman strength. The denouement of the novel is a fight between the two *marasa*, a titanic struggle between good and evil.

Stanley Péan has inscribed the power of the spirits and the supernatural in a novel that is completely based in the contemporary realities of Haitian history and the Montréal jazz scene. An aspect of the novel with which I do not have the expertise to deal is that Stanley Péan has structured his novel not only within the culture of Haitian sorcery but also within the culture of jazz, naming each unnumbered chapter after a specific work played by well-known musicians such as Theolonius Monk, Duke Ellington, Winston Marsalis, Herbie Hancock, Miles Davis, and others. The important aspect of the Vodou ambiance is that Péan posits the reality of psychokinetic relations between the *marasa*.

In his long, dramatic poem, *Un arc-en-ciel pour l'occident chrétien* (1977), René Depestre gave voice to a number of the principle Vodou spirits, who express their own strength and essence in a series of 16 "epiphanies." Dambala-Ouedo, for example,

Here I am, Dambala-Ouedo
Aquatic black man, black river
I am the beating heart of the water
I am the erect penis of the water
I wet a sprig of basil
In a glass of white wine

And I spray your wan faces
I spray your pale hysterias
I water the terror that coils up in your eyes
I water the cardinal points of your vices
I slither on my back I drag my rada
I slide I dance my yanvalou in your home
If you see a green serpent
Dance with your oldest daughter, it's me! (Translation mine.)

For Depestre, Vodou is at once "one of the active elements that enter into the formation of the national culture of the Haitian people" and "a very rich mythology that is able to give a marvelous fertility to the creative efforts of the poets and writers of Haiti" (*Arc-en-ciel* 1977, 139). I would suggest that Depestre modernized, in the mid-1960s, the revolutionary discourse of Boukman in order to inspire the Black peoples of the world to stand firm against the domination and abuse of the white, industrialized nations, with their nuclear weapons.

There are a number of novelists who have included respectful depictions of Vodou as a daily part of the existence of rural Haitians. When Manuel returns from the cane fields of Cuba to his village, in Jacques Roumain's *Gouverneurs de la rosée* (1944), his mother is duty-bound to organize a ceremony to thank the *lwa* for her son's survival and return. In Marie Chauvet's third novel, *Fonds-des-Nègres* (1960), Marie-Ange is a city girl who goes to spend the vacation with her grandmother in the *mòn* of Grande Anse. The abrupt death of her mother forces Marie-Ange to remain permanently in Fond-des-Nègres, but it is through the Vodou ceremony that she reawakens to her own culture and begins to feel that this is her home. The likelihood is that she is going to serve the *lwa* and live in Fonds-des-Nègres. Although Hilarion Hilarius, the protagonist of Jacques Stephen Alexis's first novel, *Compère Général Soleil* (1955), is condemned to a month in prison for breaking and entering into a bourgeois house in Bois Verna, he is basically an honest and intelligent young man lacking in education and a regular job. After he is freed and finds work, his mother decides that it is time to organize a ceremony of thanksgiving for the spirits. Hilarion is a city kid who resents being dragged off to the country to participate in a costly manje *lwa* for no visible return. He begins to observe and appreciate the simple life of the people, however, from the moment that a mother asks the taptap driver to stop so her son can take a leak to the point when he renews acquaintance with his ailing uncle and his cousin, who has grown from a spindly girl to an attractive young lady. During the evening ceremony, he finds himself under the influence of the peasant culture and beliefs from which he had become estranged. Jacques Stephen Alexis and Marie Chauvet both wrote their descriptions of Vodou culture and ceremonies in the light of Roumain's *Gouverneurs de la rosée*, and the ethnological research that had preceded it.

The fourth category of works is that in which Vodou and the supernatural are fundamental aspects of the novelistic world. Pétion Savain's novel, *La case de Damballah* (1939–1943), certainly came in the wake of the indigenist novels of the 1930s, but Max Dominique's accusation that Savain presents a "touristic" view of Vodou is exaggerated (1988, 143). The action of the novel is inscribed within the vision of the peasants of the mountains and high plateaus in southeastern Haiti, in the region of the villages of Bodari, Oriani, and Savann Zombi, not far from the border. Rebelné sets up housekeeping with Céline and they cultivate a garden. Rebelné acquires a champion rooster. However, misfortune falls upon Rebelné and the region. The price of coffee falls and they can no longer make a living. The serpent disappears from the dwelling of Danbala. Rebelné returns to Port-au-Prince to try to eke out an existence for himself and his family, but he succumbs to the rigors of overwork. Toward the end of the novel, however, Rebelné's mother, Sor Mée, feels that the *lwa* have punished them because Damballah's sanctuary was desecrated and abandoned. That sanctuary is restored and the gray snake that lived in it returns. This is a sign that the fortunes of the village will take a turn for the better. The novel is divided into just two parts: "Ceux d'en haut" (The people of the plateau) and "Sous l'emprise du Vodou" (Under the sway of Vodou). References to past history, the U.S. occupation and the campaign to check the Caco rebellion, and to the troubled economy, when the bottom falls out of the coffee market, are passing allusions while the world is recounted from the standpoint of the peasants of the plateau.

I would like to quote from just one passage to give a sense of the depiction of a Vodou ceremony. Rebelné leaves his work in Port-au-Prince, early in the novel, to go up to the village to see Céline. He arrives at the moment when a Vodou ceremony is beginning. The *bòkò* (the term is not used pejoratively to designate a sorcerer here), Alcinoüs, is described:

> He was a short, dirty man. Covering his large belly, he was wearing a yellow hemp beard.
>
> His drooping shoulders supported an enormous, bushy head with the powerful jaws of a stubborn old man. His expressionless face jealously hid his thoughts just as the gloomy, humid thickets of the Vault of the Black Maroons kept their mysteries.
>
> His lifeless eyes, bilious and bloodshot, had a strange aura that changed everything around him. He uttered incomprehensible words, in a husky voice, as he ran about the sanctuary with lowered head, constantly shaking his copper bell.
>
> People had brought red-bordered napkins and tablecloths that the nervous ounsi were spreading on the floor. With the place setting finished, Alcinoüs placed a handsome red rooster in the middle of the peristil and drew a vèvè around him: a large circle traced in cornmeal. Then the ounsi, dressed in red, presented the djòk to him: an earthenware dish with bits of raw sweet potato, banana, and yam that was set in front of the rooster with magic incantations. A tremendous sigh rose to the peak of the arched trees of the enclosure and reverberated for some time between the mountains. Without moving, the rooster ate in the middle of the vèvè. The lwa were content and would certainly enter the heads of the ounsi. (1939, 36–37)

Perhaps what troubled Max Dominique is that this description is so low-key, without stressing in any way the actual possession of the *bòkò* or the *ounsi*, the assisting initiates. In fact, this matter-of-fact approach strikes me as underscoring, first of all, the naturalness with which the peasants live their Vodou culture and its reality in their lives.

Martin Toma (1991), by Margaret Papillon, is a fairly recent novel. The protagonist of this novel is a young man who has been raised by exploitative, dishonest relatives. Martin Toma is finally deluded by a sharpster into buying a "passport" that will allow him to go to Port-au-Prince to look for work. Arriving in Port-au-Prince, he discovers that he has been duped—the "passport" is a false document for which he had no need. On the other hand, of course, he discovers that it is difficult to find any honest work in the vicinity of Port-au-Prince, but he begins to find odd jobs. Forced to sleep in the open at times, he discovers the secret entrance to a well-stocked and comfortable apartment in the basement of the burned Cathedral. He takes the risk of eating his fill, for the first time in a long while, and falls asleep. When the person who lives in the hidden apartment returns, Martin learns that his host is Bertrand Delbeau, who died on July 10, 1964, a little over a month prior to the "Vespers of Jérémie," when Duvalier's *makout* and the army took vengeance on the mulattoes of Jérémie for the invasion in which some of their sons had participated . By some mysterious means, Bertrand's spirit has continued to be present on the earth with a mysterious mission that has not been revealed to him at the moment when he takes Martin under his tutelage. The greater part of the novel is given over to the education of Martin, the beginning of his career as a lawyer, and his marriage. Bertrand's mission was to mentor this orphan, with his rare combination of intelligence, physical strength, and courage. Margaret Papillon's vision for Haiti goes from the year of the big emigration of Haitian intellectuals, 1964, to the year 2004, when Martin has led the revolutionary movement that establishes a true democracy in Haiti and has become the wise, disinterested counselor of the new president. At that point, Bertrand dies definitively, in a great cataclysm of cosmic force that completely destroys the semi-ruined Cathedral.

Margaret Papillon's novel is never specifically inscribed in the traditions of Vodou and Bertrand Delbeau is not presented as a *lwa*, but through the enveloping presence of the supernatural and the visionary prediction of the triumph of a true popular democracy, guided by enlightened wisdom, this fictional vision is certainly neither hostile to nor alienated from a spiritual vision of a peaceful existence in harmony with the spirits. When I spoke with Mme Papillon last fall, she told me that the idea of the novel came to her in a dream and we concurred in the fervent hope that the bicentenary of the Haitian Revolution might see a dramatic change in the Haitian state.

Lilas Desquiron's novel *Les chemins de Loco-Miroir* (1990) begins at the same moment as the novel by Margaret Papillon, just before the Vespers of Jérémie, and ends at an indeterminate future moment in the twenty-first century. This novel, however, is pervasively inscribed within the traditions of Vodou, which begins with the birth of two daughters, Cocotte to a peasant mother in the *mòn* and Violaine, to a bourgeois family in Jérémie. The *lwa* have decreed that these two girls, not related by blood, are *marasa* and both mothers accept this decree and participate in the ceremony that consecrates the special status of the *marasa* "sisters." The two girls will be raised together when Cocotte is brought down from the *mòn* to live as a *restavèk* with the Delavigne family. Violaine, the upper-crust daughter, is the twin with the nature of a wild cat and who has untameable red hair, while the peasant sister, Cocotte, is the more sensible twin, endowed with insight and with a forbearing character. The Vodou practices of the people are presented as a normal part of the culture to which all classes (bourgeois and peasant) subscribe.

Violaine falls in love with Alexandre, a young man of completely African features and, in addition, a member of a revolutionary movement that has organized an invasion to overthrow the Duvalier dictatorship. Unable to control the behavior of an impulsive daughter who rejects a proper, mulatto suitor, Violaine's mother calls a family counsel. Mother and aunts decide that this scandal cannot be allowed to destroy the family's dignity and they call in a *bòkò* to zombify Violaine. She puts on a mysterious pair of treated house slippers that evening and immediately falls into a zombified state. After the burial, the *bòkò* secretly removes Violaine's body from the isolated tomb and she is left in the care of her old nurse. The rejected, love-crazed suitor has received permission from the family council to "possess" the inanimate body of his fiancée. In scenes of virtually insane rage, he rapes the unresponsive body repeatedly, without managing to provoke the least reaction or sign of life. In revulsion at this situation, the old nurse feeds her protégée salted lamb stew and sends her wandering off into the mountains before the family can discover the act of rebellion. The wandering, semiconscious Violaine encounters a sympathetic *oungan* who takes her to a manbo. Through a healing ceremony, Violaine is brought further back toward consciousness, without regaining memory and the full use of her faculties. The *manbo* leads Violaine to Port-au-Prince, where Cocotte is working as a vendor in the Iron Market. Through this cosmic coincidence, the *marasa* are reunited and the novel ends at a future moment when Cocotte is telling the story of her life, after the death of Violaine and all other actors in the drama. The only vision of a future is that of the 80-year-old *marasa* who hopes that they have finished with both *makout* and the world of the spirits.

Although the novel is simply divided into sections narrated by one voice and, toward the end, groups of voices, it falls naturally into three parts (in keeping with the tripartite nature of the *vèvè* for the *marasa*): the first part is the history of the *marasa* sisters, Violaine and Cocotte; the second part is the passion of Violaine, who is forced to abort the child conceived out of wedlock and is finally zombified by her bourgeois family; the third part is the burial, the secret removal from the tomb, the subsequent wanderings of the partly conscious Violaine, and her eventual reunion with her *marasa* in Port-au-Prince.

The late Guy Laraque, a native of Jérémie, wrote a long and scathing review of *Les chemins de Loco-Miroir*, expressing primarily his judgment that Lilas Desquiron had composed a vicious attack on the mulattoes of Jérémie. As perceptive as Guy Laraque was, in general, I think that he was misled in taking the novel as a *roman-à-clefs*. First of all, the novel is clearly an anti-Duvalier narrative, although Lilas Desquiron's primary intention was probably not to write a "political" novel. Second, even the mulatto family of Violaine is presented, along with the peasant mother of Cocotte, within the framework of those Haitians who sincerely serve the spirits. Their crime, or sin, according to one's perspective, is their willingness to achieve a desired end by any means, including sorcery and the use of poison. After publishing an article in which I probably erred in underscoring the historical framework of the novel, I have come back to viewing the novel as a very personal work, deeply inscribed in Vodou traditions, in which the author's personal relations with the spirits must be sensed and left somewhat to the individual imagination.

I have kept Jacques Stephen Alexis's second novel, *Les arbres musiciens* (1957), as the final example for my discussion because I find in it both historical vision and a dramatic narrative of an infamous era of Haitian history, the anti-superstition campaign of 1941.

The Catholic Church had never given up in its attempts to eradicate Vodou from Haiti. The fact is, however, that President Élie Lescot was, early in his career, secretly planning to evict the peasants of certain areas in order to allow U.S. agricultural officials to carry out an experiment for cultivating rubber plants in Haiti, the *kòn kabrit* or cryptostegia, a kind of weed from which they hoped to make latex rubber for wartime use. The anti-superstition campaign, preached with new fervor by Catholic priests, brought about the destruction of some Vodou sanctuaries and served as a cover for the seizure of land from the peasants. Under the fictitious name of Pierre Roumel, as in *Compère Général Soleil*, Alexis evokes the opposition of Jacques Roumain to the anti-superstition campaign.

Alexis (1957) had undoubtedly witnessed Vodou ceremonies in his childhood since his maternal grandmother had her own *ounfò*, at Souvenance, near Gonaïves. Alexis himself says, however, that he consulted *oungan* in conjunction with the composition of *Les arbres musiciens*. This intimate view of Vodou shows through clearly. In chapter 7 of *Les arbres*, there is a great council of *oungan* at Nan Remanbrans. They have come together to consult about how to survive the new campaign against their religion. One *oungan* speaks up, saying that "Vodou is the soul of the people, its true faith and only resource." He continues, "Politics, production, commerce, industry, education, sports, culture, and the dreams of men are all influenced by the people's religion. By what right do the whites assert that their religion is the only valid religion?" (1957, 174). When Danger Dossous, a treacherous *bòkò* attempts to enter the assembly, Papa Bois-d'Orme hypnotizes him and turns him away. When Bois-d'Orme speaks to the assembled *oungan*, he foresees that their *ounfò* will be burned but this will not destroy Vodou. "The ounfò are indestructible, like the grass that is burned to the ground in order to produce a better crop. The ounfò will grow up once more during the springtime of life! Laugh at the flashing, rapid arms. Ago ye!" (1957, 179). Toward the end of the novel, Bois-d'Orme sets fire to his own sanctuary of Nan-Remanbrans. When the Reverend Diogène Osmin, who has been a leader in preaching the destruction of the Vodou temples, arrives, Bois-d'Orme addresses him in no uncertain terms:

> The lwa are immortal, priest! Look, look with your two eyes. The lwa have not allowed your sacrilegious hands to be lifted against the ancient Remanbrans. The old sanctuary is nothing but flames and ashes, and the lwa still live! Look at the light shining in the ashes, Nan Remanbrans lives! Bois-d'Orme can now march off to his death, for the former sanctuary will reemerge one day, in the same place, grander, taller, more beautiful, eternal like the lwa of eternal Africa. For your own misfortune, you will survive, but nobody will be more lifeless than you. When you look at the trees, you will see in their swaying the invisible and mystical body of the spirits. When you listen to the wind groaning over the countryside, it will be their voices cursing you. When you smell the aromas of the seasons and the perfume of the crops, you will be unable to tolerate the reproach that springs from the entrails of the earth that is no longer your earth. In the least fruit born of the soil, the essence of the lwa will seize you by the throat. Your very hands will be unable to grasp anything without the lwa devouring you with a living fire. For you have failed to respect the right of men to believe according to their own heart. Go, homeless child! Man without a race! Man without a land! Man of no nation! The hands of the spirits are upon you! (1957, 359).

* * *

Let me rapidly summarize the categories covered of the creative use of Vodou in Haitian literary works.

There are, first of all, the fairly numerous depictions of the abuse of sorcery, sometimes associated with Vodou. In this category I cited the novels by René Depestre (*Le mât de cocagne*), Franketienne (*Les affres d'un défi*), Gérard Étienne (*La Reine Soleil levée*), Stanley Péan (*Zombi Blues*), and Franck Fouché's play (*Général Baron-la-Croix*). It is probably significant, in the realms of social psychology and historical evolution, that the 30-year dictatorships of Papa Doc and son Baby Doc have provoked such a number of intense and sometimes satirical works.

Second, I mentioned Depestre's long poem, *Arc-en-ciel pour un occident chrétien*, as the only example that came to mind of a Vodou-inspired revolutionary discourse. There are numerous individual poems in honor of various *lwa* by Depestre, Morisseau-Leroi, and other Haitian poets.

With regard to Vodou presented, along with other aspects of peasant culture as an integral part of everyday life, I mentioned Jacques Roumain (*Gouverneurs de la rosée*), Jacques Stephen Alexis (*Compère Général Soleil*), and Marie Chauvet (*Fonds-des-Nègres*). The work of Roumain, along with that of Jean Price-Mars, was seminal in the ethnological materials he published and by the example he gave of incorporating Haitian folk culture into creative literature.

Finally, in a noncomprehensive survey, I gave rapid glimpses of several novels that are virtually inscribed in the Vodou vision—Pétion Savain (*La case de Damballah*), Margaret Papillon (*Martin Toma*), Lilas Desquiron (*Les chemins de Loco-Miroir*), and Jacques Stephen Alexis (*Les arbres musiciens*).

I should give credit in closing to Léon-François Hoffmann, who, in some sense, blazed a trail for subsequent readers and critics of Haitian literature with his *Le roman haïtien*. In his chapter on "Les dieux" (1982, 248–272), he summarized the presentation of Vodou in Haitian fiction by those novelists who recognized it as a legitimate religion.

> What most novelists seem to ask of the elites is not that they convert to Vodou, but to assume it, to the same extent that they assume Negritude and . . . the Kreyòl language; to recognize its beauty and poetry; to assume a degree of responsibility for its obscurantist and occasionally criminal aspects; to understand that to scorn Vodou by perpetuating the misery and alienation that nurture it is the height of insincerity. (1982, 269–270)

I have to admit that, at a moment when I had not begun working seriously in the area of Haitian literature, Hoffmann preceded me in sensing the exceptional nature of Alexis's second novel: "I think that no novelist (with the possible exception of Jacques-Stephen Alexis) was able to integrate Vodou in a coherent vision of the world" (1957, 270).

In conclusion, I note that to my knowledge (and Hoffmann mentions no examples either), there have been no Haitian novelists who were themselves *sèvitè* of the *lwa*, none who have admitted it, at least. Jacques Roumain, Jacques Stephen Alexis, and the still living novelist, Lilas Desquiron, are among those who were most knowledgeable and the most careful to offer serious depictions of Vodou and to separate the religion from its abusive use by heads of state or sorcerers. Among the effects of creative presentations of Vodou, it is necessary to recognize that the language, and the culture, have been enriched by the terminology of Vodou and the names of its spirits.

A Hasty Glossary of Vodou and Other Cultural Terms in Kreyòl

bòkò: a priest who practices sorcery; sometimes used in earlier literature as a synonym for an *oungan*.

dechoukaj: the ouster of a Haitian president by a popular revolt; Baby Doc was the target of the most recent *dechoukaj*.

kòn kabrit: cryptostegia americensis—a species of weed that the United States introduced into Haiti in 1941–1942, hoping to develop an alternate source of latex for rubber; the project was a dismal failure.

lwa: spirit(s) of Vodou.

manbo: a Vodou priestess.

manje lwa: a ritual meal for the spirits.

marasa: twins in Vodou; the twins are highly honored, but they are capable of great mischief if they feel that their importance has not been properly recognized at a ceremony or a ritual meal. The *dossou/dossa* (male/female child born after the twins) is also highly respected and figures in the *vèvè* of the marasa (see example).

mòn: mountain(s), hill(s).

ounfò: a sanctuary.

oungan: a Vodou priest.

restavèk: a child who is taken into virtually indentured service by a well-to-do family in Haiti; many poor parents have felt forced to place children as *restavèk*; the children may be relatively well treated but are never (to my knowledge) given the same privileges as the children of the family and there are many attested cases of abuse. The practice still continues, in spite of much adverse publicity.

tonton makout: member(s) of the François Duvalier's militia, the VSN (Volontaires de la Sécurité Nationale); authorized to use any form of violence at their own discretion.

vèvè: the symbol of specific *lwa*, drawn on the ground with coffee grounds, flour, and corn meal before the beginning of a ceremony.

zombi: a person whose soul has been captured by a sorcerer, leaving the individual without a will of their own.

Bibliography on Vodou

Fiction

Alexis, Jacques Stephen. 1955. *Compère Général Soleil.* Paris: Gallimard.
Translation: *General Sun, My Brother.* Intro. and trans. Carrol F. Coates. University Press of Virginia, CARAF Books (2005).

Translation: *General Sun, My Brother.* 1957. *Les arbres musiciens.* Paris: Gallimard.
———. 1960. *Romancero aux étoiles* [short stories]. Paris: Gallimard.
Translation: "The Enchanted Second Lieutenant" ("Le sous-lieutenant enchanté"). Trans. Sharon Masingale Bell, *Callaloo,* 20 (3) (Summer 1997): 504–515.
Carpentier, Alejo. 1989. *El reino de este mundo.* New York: Farrar, Strauss and Giroux.
Translation: *The Kingdom of This World.* Trans. Harriet de Onis. New York: Noonday Press/Farrar, Straus and Giroux, 1989.
Casséus, Maurice. *Mambo,* ou *la Revanche des collines.* Port-au-Prince.
Chauvet, Marie. 1960. *Fonds-des-Nègres.* Port-au-Prince: Henri Deschamps.
Translation: Excerpt from *Fonds-des-Nègres.* Trans. Carrol F. Coates with notes, *Callaloo,* 14 (3) (Summer 1991): 564–568.
Chevallier, André, and Luc Grimard. 1950. *Bakoulou.* Port-au-Prince: Soc. d'Édition et de Librairie.
Colimon Hall, Marie-Thérèse. 1974. *Fils de misère.* Port-au-Prince: Éds. des Caraïbes.
Dartiguenave, Edith. n.d. *L'héroïsme d'une Haïtienne.* Port-au-Prince: [sans éditeur].
Defay, Louis. 1953. *Ceux de Bois-Patate.* Port-au-Prince: Impr. de l'État.
Depestre, René. 1988. *Hadriana dans tous mes rêves.* Paris: Gallimard.
Translation: Excerpt from *Hadriana in All My Dreams.* Carrol F. Coates, *Callaloo,* 15 (2) (Summer 1992): 546–549.
———. 1979. *Le mât de cocagne.* Paris: Gallimard.
Translation: *The Festival of the Greasy Pole.* Trans. and intro. Carrol F. Coates. University Press of Virginia.
Desquiron, Lilas. 1990. *Les chemins de Loco-Miroir.* Paris: Stock.
Translation: *Reflections of Loko Miwa.* Intro. Marie-Agnès Sourieau; trans. Robin Orr Bodkin. University Press of Virginia, "CARAF Books," 1998.
Étienne, Gérard. 1979. *Un ambassadeur macoute à Montréal.* Montréal: Nouvelle Optique.
———. 1983. *Une femme muette.* Montréal: Nouvelle Optique.
———. 1988. *La Reine Soleil levée.* Montréal: Guérin Littérature.
Fignolé, Jean-Claude. 1987. *Les possédés de la pleine lune.* Paris: Seuil.
Frankétienne. 1979. *Les affres d'un défi.* Port-au-Prince: Henri Deschamps.
Translations: "Defiance and Dread" [beginning of novel]. Trans. J. Michael Dash. *Callaloo* 15 (2) (Spring 1992): 393–397; excerpt from "*Les affres d'un défi*" [the following section]. Trans. Carrol F. Coates with a note, *Callaloo,* 19 (3) (Summer 1996): 756–761.
Godart, Jacques. 1987. *Pourquoi les campêches saignent-ils?.* Port-au-Prince: Henri Deschamps.
Greene, Graham. 1966. *The Comedians.* London/NY: Penguin Books.
Innocent, Antoine. 1906. *Mimola* ou *L'histoire d'une cassette. Petit tableau de moeurs locales.* Port-au-Prince: V. Valcin.
Laferrière, Dany. 1997. *La Chair du maître* [short stories]. Montréal: Lanctôt.
Translation: "A Country Wedding" ("Un mariage à la campagne"). Trans. Carrol F. Coates, *Callaloo* 22 (4) (Autumn 1999): 900–909.
Léon-François Hoffmann. 1982. *Le roman haïtien. Idéologie et structure.* Sherbrooke: Éditions Naaman.
Lesage, Aimard [Desaix Baker]. 1926. *L'Amant idéal.* Port-au-Prince: Impr. Chéraquit.
Papillon, Margaret. 1991. *Martin Toma.* Port-au-Prince: L'Imprimeur II.
Péan, Stanley. 1996. *Zombi Blues.* Montréal: La Courte Échelle.
Pierre, Romulus. 1978. *Les Zombis en furie.* Port-au-Prince: Fardin.
Roumain, Jacques. 1944. *Gouverneurs de la rosée.* Paris: Éditeurs Français Réunis.
Translation: *Masters of the Dew.* Intro. J. Michael Dash, trans. Langston Hughes and Mercer Cook. London: Heineman, 1978.

———. 1931/1987. *La montagne ensorcelée*, récit paysan in *La montagne encorcelée* [recueil d'écrits en prose et en vers], 52–116. Paris: Messidor.
Savain, Pétion. 1939/1943. *La case de Damballah.* Port-au-Prince: Impr. de l'État; 2e éd., Paris: Les Oeuvres Nouvelles; repr., Heinemann, 1970.
Thoby-Marcelin, Philippe and Pierre Marcelin. 1944. *Canapé-vert.* New York: Éds. de la Maison Française.
———. *Canapé Vert.* Trans. Leonart Thomas. New York: Farrar and Rinehart, 1944.
———. 1980. *Tous les hommes sont fous.* Montréal: Nouvelle Optique.

Poetry

Depestre, René. 1977. *Arc-en-ciel pour un occident chrétien, poème-mystère vaudou.* Paris: Présence Africaine.
Translation: *Rainbow for the Christian West.* Trans. Joan Dayan. Boston: University of Massachusetts Press.

Theater

Fouché, Franck. Général Baron-la-Croix ou le Silence masqué. Tragédie moderne en 2 calvaires, 28 stations et une messe en noir et rouge. In *Francophonie Vivante.* Montréal, Leméac.
Translation: Excerpt from *Général Baron-la-Croix . . .* Trans. Carrol F. Coates, *Callaloo* 15 (3) (Summer 1992): 680–686.
Vieux, Isnardin. 1974. *Mackendal, drame historique.* Port-au-Prince: Imprimerie Centrale.

Studies

Antonio, Louis-Jean. 1970. *La crise de possession et la possession dramatique.* Montréal: Leméac.
Apollon, Willy. 1976. *Le Vaudou, un espace pour les "voix."* Paris: Éds. Galilée.
Bastide, Roger. 1967. *Les Amériques noires.* Paris: Presses Universitaires de France.
———. *African Civilizations in the New World.* Trans. Peter Green. New York/London, Harper and Row.
Bellegarde-Smith, Patrick. 1990. *Haiti. The Breached Citadel.* Boulder, CO: Westview Press; 2nd ed., Toronto: Canadian Scholars Press, 2004.
Benoît, Max, ed. 1978? *Cahier de folklore et des traditions orales d'Haïti.* Port-au-Prince, Impr. La Phalange. (Mémoires présentés par des élèves du Prof. Max Benoît.)
Bijoux, Legrand. 1990. *Coup d'oeil sur la famille haïtienne.* Port-au-Prince, Éditions des Antilles.
Brodwin, Paul. 1996. *Medicine and Morality in Haiti: The Contest for Healing Power.* Cambridge University Press.
Brown, Karen McCarthy. 1991. *Mama Lola. A Vodou Priestess in Brooklyn.* Berkeley: University of California Press; 2nd ed., Berkeley: University of California Press, 1994.
Cabon, P. A. 1933. *Notes sur l'histoire religieuse d'Haïti. De la Révolution au Concordat (1789–1860).* Port-au-Prince: Petit Séminaire, Collège Saint-Martial.
Caplain, Jules. *La France en Haïti. Catholicisme, vaudoux, maçonnerie.* Paris.
Comhaire-Sylvain, Suzanne. 1938. *A propos du vocabulaire des croyances paysannes.* Port-au-Prince.

Cosentino, Donald J., ed. 1995. *Sacred Arts of Haitian Vodou*. Los Angeles, CA: UCLA Fowler Museum of Cultural History.
Courlander, Harold. 1939. *Haïti Singing*. Chapel Hill: University of North Carolina Press.
———. 1960. *The Drum and the Hoe. Life and Lore of the Haitian People*. Berkeley: University of California Press.
Courlander, Harold, and Rémy Bastien. 1966. *Religion and Politics in Haiti. Two Essays by H.C. and R. B.* ["Vodoun in Haitian Culture," by Harold Courlander; "Vodoun and Politics in Haiti," by Rémy Bastien]. Washington, DC: Institute for Cross-Cultural Research.
Dash, J. Michael. 1981. *Literature and Ideology in Haiti, 1915–1961*. NY: Barnes and Noble Books.
Davis, Wade. 1986. *The Serpent and the Rainbow*. London: Harper Collins.
Dauphin, Claude. 1986. *Musique du Vaudou. Fonctions, structures et styles*. Sherbrooke: Naaman.
Dayan, Joan. 1995. *Haiti, History, and the Gods*. Berkley: University of California Press.
Déita [Mercédes Foucard Guignard]. 1993. *La légende des loa, Vodou haïtien*. Port-au-Prince: [sans éditeur].
Delbeau, Col. Jean-Claude. 1990. *Société, culture et médecine populaire traditionnelle. Étude sur le terrain d'un cas: Haïti*. Port-au-Prince: Henri Deschamps.
Deren, Maya. 1949. *Divine Horsemen. The Living Gods of Haiti*. London and NewYork: Thames and Hudson.
Desmangles, Leslie G. 1992. *The Faces of the Gods. Vodou and Roman Catholicism in Haiti*. Chapel Hill: University of North Carolina Press.
Desquiron, Lilas. 1990. *Racines du Vodou*. Port-au-Prince: Henri Deschamps.
Dominique, Max. 1988. *L'arme de la critique littéraire: litérature et idéologie en Haïti*. Eds. Montréal: CIDIHCA. [Two chapters are particularly relevant: chapter 7, "Questionner *Dezafi*," on Franketienne, and chapter 8, "Vodou et littérature."]
Dorsainvil, J.-C. 1931. *Vodou et névroses*. Port-au-Prince: Impr. La Presse.
———. 1937. *Psychologie haïtienne. Voudou et magie*. Port-au-Prince.
Dunham, Katherine. 1969. *Island Possessed*. Chicago: University of Chicago Press.
Fleurant, Gerdes. 1996. *Dancing Spirits. Rhythms and Rituals of Haitian Vodun, the Rada Rite*. Westport, CT: Greenwood Press.
Fouchard, Jean. 1953. *Les marrons du syllabaire*. Port-au-Prince: Henri Deschamps.
Fouché, Franck. 1976. *Vaudou et théâtre: pour un nouveau théâtre populaire*. Montréal: Leméac.
Foubert, Alain. 1990. *Forgerons du Vodou*. Port-au-Prince: Deschamps.
Greene, Anne. 1993. *The Catholic Church in Haiti. Political and Social Change*. East Lansing, MI: Michigan State University Press.
Herskovits, Melville J. 1937. *Life in a Haitian Valley*. New York: Alfred A. Knopf.
Hoffmann, Léon-François. 1982. *Le roman haïtien. Idéologie et structure*. Sherbrooke, Canada: Éditions Naaman.
———. *Haïti: couleurs, croyances, créole*. Port-au-Prince/Montréal, Henri Deschamps/Éds. CIDIHCA.
Holly, Arthur. 1918. *Les Daimons du culte vodou*. Port-au-Prince, Edmond Chenet.
Honorat, Michel Lamartinière. 1955. *Les Danses folkloriques haïtiennes*. Port-au-Prince, Pubs. du Bureau d'Ethnologie de la République d'Haïti, Sér. 2, No. 11.
Hurbon, Laënnec. 1987. *Le barbare imaginaire*. Port-au-Prince: Henri Deschamps.
———. 1987. *Comprendre Haïti. Essai sur l'État, la nation, la culture*. Port-au-Prince: Henri Deschamps.
———. 1987. *Culture et dictature en Haïti. L'imaginaire sous contrôle*. Port-au-Prince: Henri Deschamps.
———. 1987. *Dieu dans le vaudou haïtien*. Port-au-Prince: Henri Deschamps.

———. 1989. *Le phénomène religieux dans la Caraïbe. Guadeloupe—Martinique—Guyane—Haïti.* Montréal: Éds. CIDIHCA.
———. 1993. *Les mystères du vaudou.* France: Gallimard "Découvertes."
———. *Voodoo. Search for the Spirit.* Trans. Lori Frankel. New York: Harry N. Abrams.
Hurston, Zora Neale. 1938. *Tell My Horse. Voodoo and Life in Haiti and Jamaica.* New York: Lippincott.
Kerboull, Jean. 1973. *Le Vaudou.* Paris: Robert Laffont.
Kergoz, Jean. 1921. *L'expulsion des Vaudoux.* Paris: Lib. Générale Cahtolique Arthur Savaete.
Laguerre, Michel. 1979. *Études sur le Vodou haïtien.* Montréal: Université de Montréal.
———. 1980. *Voodoo Heritage.* Beverly Hills, CA: Sage Publications.
———. 1989. *Voodoo and Politics.* New York: Saint Martins.
Lanternari, V. 1962. *Les mouvements religieux de liberté et de salut des peuples opprimés.* Paris: Maspéro.
Laroche, Maximilien. 1991. Musique, Danse, Religion. In *La double scène de la représentation. Oraliture et littérature dans la Caraïbe,* 85–114. Collection, "Essais," no. 8. Université Laval, GRELCA.
Léveque, Karl. 1971. L'interpellation mystique dans le discours duvalierien. *Nouvelle Optique,* December: 5–32.
Louis-Jean, Antonio. 1970. *La crise de possession et la possession dramatique.* Montréal: Leméac.
Marcelin, Milo. 1949–1950. *Mythologie vaudou.* 2 vols. Port-au-Prince.
Mars, Louis. 1946. *La crise de possession dans la Vaudou. Essai de psychiatrie comparée.* Port au-Prince: Impr. de l'État.
Maximilien, Louis. 1945. *Le Vaudou haïtien. rite Radas—Canzo.* Préf. du Dr. Pierre Mabille. Port-au-Prince: Impr. de l'État.
McAlister, Elizabeth, ed. 1996. *Angels in the Mirror. Vodou Music of Haiti.* Roslyn, NY: Ellipsis Arts (includes CD).
Mennesson-Rigaud, Odette. 1958. Le rôle du Vaudou dans l'indépendance d'Haïti. *Présence Africaine,* February–March, 17 and 18: 43–67.
Métraux, Alfred. 1958. *Le Vaudou haïtien.* Paris: Gallimard.
———. *Voodoo in Haiti.* Trans. Sidney W. Mintz. New York: Pantheon, 1989.
Michel, Claudine. 1995. *Aspects éducatifs et moraux du Vodou haïtien.* Port-au-Prince: Le Natal.
Montilus, Guérin. 1968. *Les repas sacrés haïtiens. Essai d'interprétation théologique.* Paris.
Moreau de Saint-Méry. 1954. *Description topographique, physique, civile, politique et historique de la Partie Française de l'Isle Saint-Domingue.* 3 vols. Nouv. éd. par Blanche Maurel et Étienne Taillemite. Paris: Société de l'Histoire des Colonies Françaises.
Murray, Gérald. 1980. *Population Pressure, Land Tenure, and Voodoo.* New York: Academic Press.
Nérestant, Micial M. 1994. *Religions et politique en Haïti (1804–1990).* Préface de Louis de Vaucelles. 2nd ed. Paris: Éds. Karthala.
Peters, Carl Edward. 1941. *Lumière sur le houmfort.* Port-au-Prince: Cheraquit.
———. 1956. *Le service des "loas."* Port-au-Prince: Impr. Telhomme.
Planson, Claude. 1974. *Vaudou: Un initié parle.* Paris: Jean Dullis.
———. 1987. *Le Vaudou.* Paris: MA Editions.
Pressoir, Ch.-Fernand. 1974. *Débats sur le créole et le folklore.* Port-au-Prince: Impr. de l'État.
Price-Mars, Jean. 1998. *Ainsi parla l'oncle.* Nouv. éd. Montréal: Leméac. Repr. Port-au-Prince: L'Imprimeur II.
———. *So Spoke the Uncle. Ainsi Parla l'Oncle.* Trans. and intro. Magdaline W. Shannon. New York: Passeggiata Press, 1997.
Rigaud, Milo. 1953. *La tradition vaudou et le Vaudou haïtien. Son temple, ses mystères, sa magie.* Paris: Niclaus.
———. 1953. *Secrets of Voodoo.* Trans. Robert B. Cross. San Francisco, CA: City Lights Books.

Rigaud, Milo. 1974. *Vè-Vè. Diagrammes rituels du Voudou.* Trilingual edition French-English-Spanish. New York: French and European Publications.
Roumain, Jacques. 1942. *A propos de la campagne "anti-superstitieuse."* Port-au-Prince: Impr. de l'État.
———. 1943. *Le sacrifice du tambour-assotor.* Pub. du Bureau d'Ethnologie de la République d'Haïti, No. 1. Port-au-Prince: Impr. de l'État.
Saint-Gérard, Dr. Yves. 1992. *Le phénomène Zombi: La présence en Haïti de sujets en état de non-être.* Toulouse: Éds. Érès.
Salgado, J. M. 1962. *Le culte africain du Vaudou et les baptisés en Haïti: Essai de Pastorale.* Rome.
Seabrook, William. 1929. *The Magic Island.* New York: Lancer Books.
Sterlin, Philippe. 1954. *Vèvès vodou.* Série I, II. Port-au-Prince: Philippe Sterlin.
Trouillot, D. 1885. *Esquisse ethnographique. Le Vodoun. Aperçu historique et évolution.* Port-au-Prince: Impr. R. Éthéart.
Verger, Pierre. 1957. *Notes sur le culte des Orisa et Vodun.* Dakar: Ifan.
Verschueren, J. *La République d'Haïti.* T. III: *Le culte du Vaudou.* Paris.
Wexler, Anna. Clotaire Bazile. 1997. Oungan and Haitian Vodou Flagmaker. (The Flags of Clotaire Bazile. A Description; An Interview with Clotaire Bazile). *Callaloo* 20 (2) (Spring): 371–398.
Williams, Joseph J. 1989. *Voodoos and Obeahs: Phases of West Indian Witchcraft.* New York: AMS Press.
Wirkus, F., and T. Dudley. 1932. *Le roi blanc de la Gonave: Le culte du Vaudou en Haïti, 1915–1929.* Paris: Payot.

Chapter 12

It's All for You, Sen Jak!

Donald J. Cosentino

Plaine-du-Nord is an old colonial town at the crossroads of a fertile plain due south of Cap Haitien. The first slave uprisings of the Revolution began on that plain in August, 1791. Henri Christophe had the Citadel built nearby, incongruously outsized, overbearing and magnificent, like the genius who envisioned it as Haiti's final redoubt against revanchist France. Plaine-du-Nord boasts no such architectural wonder, except perhaps for the church of St. James, whose plain Jesuit face is painted a pretty lemon yellow. A remnant of slavery days, the church still commands the town center. To the right of the church is the rectory where Fr. Keweillant, the Breton pastor, lives.[1] To the left, the town cemetery is dominated by a tall iron cross erected on a concrete plinth.

Running past the church and cemetery is Centenary Road, a dirt track edged by several large silk-cotton *mapou* trees. A few hundred yards beyond the church the track swerves around a series of potholes which are usually little more than large ruts.[2] Sometimes during the early summer rains these ruts fill in to become a kind of small pond called *trou* or *basin.* Should the rains fail, townspeople will come with pails of water to insure plenty of mud. For them, these are not potholes but St. James's own pond, known as the Trou Sen Jak. Its celestial sludge, along with the church, the cemetery and the surrounding countryside, mark the terrestrial emergence point for a saint who is generalissimo of a military family of spirits named Ogou. Plaine-du-Nord is certainly his most important shrine in Haiti; perhaps his most important in the whole wide Afro-Atlantic world where Ogou's avatars prevail.

Ogou ceremonies are held at (and in) the Trou most Thursdays of the year,[3] except during the month of November and throughout the season of Lent. For three days prior to the canonical feast of St. James (July 25), pilgrims from all over Haiti and its diaspora descend on the town by truck, car and foot. Trou-side ceremonies continue day and night, non-stop, for 72 hours until the town's tiny elite gathers at the Church for the official point of the festivities, the Mass for St. James, celebrated by the diocesan priests and visiting hierarchs. But by this time most of the pilgrims have departed.

They have already hopped onto tap-taps for the short ride to nearby Limonade where festivities for St. Anne—mother of Mary, grandmother of Jesus and great aunt of John the Baptist and St. James—are already underway.

Pilgrims arrive at Plaine-du-Nord wearing blue suits and red scarves, or the multi-striped *rad penitans*.[4] More secular visitors wear T-shirts celebrating the lascivious possibilities of *Getting Naked at Panama Beach*. All have come with the usual devout intentions: to fulfill vows, to seek healings, to have a good time. Pregnant women and tubercular children line up before zinc basins for a bath and a blessing from itinerant herbalists. An ecstatic woman, declaiming in a deep male voice, rides a bony bull. Set like an altar with a red ribbon around his neck, and burning candles fixed to his horns, the bull becomes a lumbering sacrifice for Ogou. After the muddy tauricide, eager pilgrims line up to be anointed by his blood.

At this festival all is synchronic. The bull falls to his knees. Nearby, pilgrims, overcome by emotion and spirit, fling themselves into the mud and lie face down, not visibly breathing. They arise, looking like primal creatures. Bystanders are moved to offer alms to beggars in the mud. Three groups of drummers playing Petwo rhythms are situated at crosspoints around the Trou. Day into night, their music never stops. Fr. Keweillant, wearing a panama hat, strolls around the Trou, joking with parishioners and explaining to inquiring visitors, "What can we do? I have always allowed this."

A penitent son of the Iron God wanders by with razor blades and safety pins stuck up and down his bare back. A woman with no legs sits near the Trou jiggling a couple of fornicating puppets. Chickens are whirled over supplicants by an oungan with one arm and no teeth. Trucks arrive with more pilgrims. Visitors gather near the *mapou*: off-beat tourists, edgy journalists, enthusiastic folklorists, would-be anthropologists; everyone taking photos, gazing, stupefied by this god's plenty. Pickpockets circle everyone, striking like sharks.

This pilgrim's progress is not limited to the Trou. Groups steadily enter the cemetery where familiar smells of burning candles and rum arise from the cross on the plinth. An oungan is pulling along a black goat, a propitiation for Bawon Samdi. Next to the cemetery an enormous crowd has gathered on the steps of the church. But they cannot enter. Iron gratings bar the doors and windows. So the pilgrims shout prayers and hurl objects: candles, pennies, cigars, rum bottles, through the gratings. They aim their missiles at an empty niche that used to contain an image of St. James. Fr. Keweillant gives specific reasons for keeping the fervent pilgrims away from their sacred target. "The Catholic Church has nothing to do with this yearly pilgrimage. These rites are for another religion. They go in front of the Church because Vodou is a very synthetic religion. It searches to reconcile divinity. These Vodouists think that an aspect of divinity is found inside the church. They call him Sen Jak, but they mean Ogou. For them, Ogou is not a saint. He is a divinity."[5]

Keweillant further explains why he has kept the doors of the parish church closed against this festival since 1978. "There were constant incidents. Perpetual sacrifices. People had immodest habits. I saw a woman lift up her skirt in front of the saint's statue and say, 'Sen Jak, here I am. It's all for you!' Another woman offered Sen Jak a piece of soap to wash her crotch, forgive me! I heard a woman in the dark part of the church say, 'Sen Jak, you are a big, powerful man. The man I live with is too old. His *zozo* (prick) doesn't work. Help me to find a younger one. . . .' I heard these sorts of things, and decided to shut the church during the pilgrimage."

How Deep is the Mud . . . ?

Keweillant was merely pointing out what every Vodouist knows. Sen Jak *is* Ogou, senior brother of a military lineage broad enough to include the founding heroes of Haitian history: Dessalines, Louverture, Christophe. And their paradigmatic struggles now encompass Aristide's Lavalas movement assembled under the standard of Ogou's fighting red cock. Through fantastic processes of appropriation, major events in Haitian history have thus been refigured through the lineaments of a Spanish warrior saint. There is no doubt that the saint's popularity rests on a perceived iconic correspondence between him and the Fon/Yoruba deity, Gu/Ogun.[6] Nor that this correspondence has inspired a creolized theology which remains profoundly affected by a continuous reinterpretation of imposed iconography. The Ogou metaphor long ago found its correlate in Catholic popular art, but where and how was the correspondence perceived? How deep, we ask, *is* the mud at Plaine-du-Nord?

Measured in time, the mud is deep indeed, sinking into the Gospel narratives and the Golden Legends which followed. Jesus nicknamed his impetuous cousin James (in French, Jacques) the "Son of Thunder." Along with Simon Peter and his brother John (the Beloved), James was closest to Jesus, being present at the Transfiguration on Mt. Tabor, and the Agony in the Garden of Gethsemane. From the Acts of the Apostles, we learn that James was martyred during Herod's first persecution. Thereafter legend takes over. According to hagiography, this same "Thunderer" traveled to Spain before his martyrdom, whence his bones were miraculously translated and discovered at Compostela in the ninth century. Out of that legend emerged the famous apparitions of Santiago (St. James) on his white horse, brandishing a flaming sword and inspiring the war against the Moors during the Iberian *reconquista.*[7]

The apparition has inspired this key image: Sword and eyes raised heavenward, Santiago rides with sublime indifference over the maimed body parts of the Moors lying trampled beneath the hooves of his equally oblivious white stallion. He is dressed all in blue, with gold scallops decorating his collar and a red medallion embossed on his wide cap. A knight mounted on a brown steed rides by his side. He is in full armor, with the casque of his helmet closed. In a mailed fist, the knight carries a white flag with a red cross.

Vodouists salute this polychrome image as a representation of the senior brother of a lineage of military spirits, all of whom share the name Ogou. "Representation" is, of course, a very pale word for an image whose properties directly reflect divinity. Mechanically reproduced by the millions, these lithographs are none the less sacred, in the same way that millions of St. Christopher medals, or knickknacks stamped with the image of the Virgin of Guadalupe, are sacred in popular Catholicism. Their commonness does not render them common, nor their ubiquity vulgar. Rather their availability is a sort of miracle, a proof of the divine presence encoded in the mysteries of lithography. And like miracles everywhere, this image has been subject to an intense scrutiny, a folk hermeneutic whose exegeses have significantly reshaped an old West African god.

Image replication is a multi-media affair: lithographs, flags, plaster, concrete and wooden statues, friezes, murals, medals, key fobs, costumes. In these various forms the Saint adorns taxis, ounfòs, and home altars in Plaine-du-Nord, Bizonton, Brooklyn, and all the other towns and places Haitians have made their own. Wherever they are, so too are lithographs of Santiago/St. Jacques/Sen Jak plastered above bedsteads and within *Rogatoires* (home altars). Magnified copies of the same image are painted onto ounfò walls, cast into huge concrete representations dominating the shrine rooms in Carrefour and Croix-des-Missions, shaped out of papier maché for collectors of Haitian art, and painted into tableaux by Vodou artists. Befitting his role as general of the armies, Sen Jak's image is appliquéd or sequined onto more Vodou *drapo* than any other lwa, except perhaps for that of his consort Ezili, with whom he is said to march in battle.

As with so many other imported objects, Vodouists have further subjected the lithograph to metonymic elaborations. Because Ogou is the lwa of iron, he is also patron of roads. Thus Haitian taxi drivers tie red ribbons around the base of their rear view mirrors;[8] or in more flamboyant obeisance, paint the image of Sylvester Stallone as Rambo on the back of tap-taps. Or sometimes Ogou is realized in the aggrandizing of his horse, as in the carousel sculpture created by Pierrot Barra. Indeed, for Barra the Ogou metaphor has few referential boundaries. To update iconography for the annual pilgrimage to Plaine-du-Nord, he created a sculpture of an Ogou airplane, painted blue and red, with images of Sen Jak and *kouzen Jezi* (Jesus) peeking out the windows.[9]

There is a powerful resonance between these Haitian manifestations of Ogou and the use of borrowed imagery in contemporary Western art. In both cases artists are appropriating commercially distributed images to forge new cultural symbols. But it took Andy Warhol to teach New York artists what the masters of the Port-au-Prince iron market had already found out: "if you blow a plain image up to absurd proportions, or reproduce it often enough, you are not sneering at its ordinariness but somehow gilding it with a glamour and pathos of its own."[10] Vicki Goldberg further explicates the process:

> For centuries, artists provided forms for society's heroes. . . . Photography, film, newsreels and television changed that, and artists are now reduced to playing with images already established in public fantasies, stoked by forces beyond the limits of art. . . . This is evidence of global envy. . . . Who is there to look up to?[11]

When the referent is invoked—"Who is there to look up to?"—Vodou and secular artists part company. For behind each divine irruption, no matter how materially fantastic, there is a reality more specific than the celluloid or plastic of its fabrication. Ogou's reality is palpable, encountered in the Trou Sen Jak, installed at the Palais National, or manifested just around the corner—at the neighborhood peristyle, where Sen Jak is riding his horse tonight.

Of Gubasas and Sangamentos

It is crucial to note that for all its outsized celebration by Haitians, the image of St. Jacques was encountered by Africans well before the first slave ever departed for the Americas. In 1480, the Portuguese built a chapel to Santiago in Elmina on the Fante coast. Thus the image of the Saint was established at the point of forced embarkation for future generations of Africans more than a decade before the crew of the Santa María ever set foot on Hispaniola.[12] The effects of this imposition on the West African imagination may be conjectured, inter alia, from iconic puns on the Christian Portuguese in Benin bronze sculpture, or in the curious parallels between the lithograph of St. James and Fon martial display as both blend into the Vodou aesthetic. As Robert Farris Thompson observes:

> The vision of killing by iron observed within the lithograph distinctly fits the martial paradigm of Dahomean Gu. And just as the *gubasa* is the central sign of war and smithing in Dahomey, so the saber became the chief icon of Ogun Ferraille in Haiti, in a new and wondrous context, often flanked by honorific banners, symmetrically displayed and inclined, as if nodding in honor of Lord Gu. . . . The shape of the shafts of the important flags that traditionally flank the sword of the saint in ground-paintings (and the sword of the master-of-ceremonies in vodun dancing) extend a little-noted accent of militaristic assertion. The flags' shafts reverse the S-curved saber's handguard; they are cryptic swords of cloth, following and flanking the lord of the cutting edge, even as a white cross on a field of red in a chromolithograph accompanied the warrior saint.[13]

As suggestive as these West African references are, Santiago made a far more portentous footfall further down the coast. In 1491, Portuguese missionaries presented a flag embroidered with the Cross of St. James to Nzinga Nkuvu, King of the Kongo.[14] For three centuries thereafter, Santiago played a decisive role in the transformation of Kongo culture, a role whose dimensions are only now being properly described by historians.[15] A letter dated from 1512 describes the Saint's part in the victory won by Dom Afonso in one of the myriad civil wars He and his propagators inspired:

> We called upon the blessed apostle Saint James, and as a result, miraculously, we saw all our enemies turn their backs and flee as fast as they could . . . after the victory we learned that those who escaped declared as one man that when we invoked the apostle Saint James they all beheld a white cross amid a great number of armed and mounted men which caused them such great dismay that they could not refrain from fleeing at once.[16]

After his victory, Afonso I combined the annual military ceremony commemorating his seizure of power with the canonical feast of St. James on 25 July.[17] De Lucques, a Portuguese observer, described one such celebration which lasted for eight days (two Kongo weeks) in the province of Soyo:

> On the feast of the saint, before coming to church to hear Mass the count goes to see the countess, from whom he receives the bow and arrows. In the presence of his wife, who sits on the ground, and together with two of his principal chiefs . . . he immediately begins to perform a private *sangamento*, this is, to leap about and flourish the bow.

> After this military drill, he returns to his palace and attends Mass, giving satisfaction first to the devil and then to God. . . . On another day there is a public *sangamento* outside the church; to see so many Blacks shouting and screaming, you would think you were witnessing a scene from Hell.[18]

De Lucques's denunciation of St. James's feast at Soyo is, curiously, not very different from the complaints of Fr. Keweillant, three centuries and one ocean removed. Both perceived bifurcated celebrations for a bifurcated saint, divine attaché for at least two heavens. Both feasts took place outside the Church, "satisfying the [African] devil," before a tiny elite went inside to satisfy the Christian God. And both engaged the resigned wrath of Catholic clergy for a religious ritual they couldn't control. The feast at Soyo, documented three hundred years ago, is in fact a process model for what Africans would reinvent in Haiti, a harbinger of the immense influence Kongo would have in Hispaniola.[19]

Processes of survival or of reinvention which have settled the image of St. James at the core of Vodou sacred art can only be conjectured from a haphazard historical record. But religious developments in St. Domingue were certainly synchronous with those in the Africa. While the cult of St. James was being propagated in Kongo, the Pope declared him patron of Hispaniola. Church records indicate the slaves embraced his cult with great enthusiasm. Several of the earliest churches built in St. Domingue bore his name,[20] though none has so consistently reflected such furious devotion to the saint as his church at Plaine-du-Nord.

Plaster representations of the saint have filled a niche in that church from the eighteenth century until his deposition and the installation of iron grates a generation ago. During all that long time the presiding clergy have continued to take a dim view of the popular devotions inspired by the shrine they reluctantly maintained. Writing for a church bulletin in 1950, Msgr. Jan noted, "The blacks of Plaine-du-Nord enjoy the satisfaction of combining in the same place the two sanctuaries of their religious beliefs, the church and the ounfò. Diocesan priests offered even sharper comments, 'I don't like to assist at the feast of St. James. I am not pleased to find myself mixed up with the idolators of Papa Ogou, and to find, en route, peristyles of dancers from whom escape the hysterical cries of those possessed by "angels." In jostling the numerous crowd, I ask myself, "Where are the Catholics? Where are the Vodouists?"[21]

The cause for all this untoward enthusiasm was not one, but two statues of St. Jacques. Marc Pean has outlined their history, which links the 1986 celebrations already described to a parish scandal in 1899. One might well ponder the history of that affair, for it foreshadows all the later attempts to "root out superstition" at Plaine-du-Nord. These fin-de-siècle events were to play out again in the *dechoukaj* which followed the overthrow of Duvalier *fils*. In Spring of 1986, an army of fundamentalist Christians, incited by the anti-Vodou broadcasts of the Protestant Radio Lumière, marched to the Trou Sen Jak. They planned to seal Ogou away forever by pouring concrete into his muddy hole. Their intentions were foiled by the drawn machetes and *cocomacaques* (weapon-sticks) of furious Vodouists who protected Ogou's point of emergence with the same resolve their ancestors showed against the iconoclast assault of Cincinnatus Leconte and his Anti-Vodou League a century before.[22]

Vodou in the Age of Mechanical Reproductions

A passion for the images of St. James has hardly been confined to Plaine-du-Nord. A larger than life-size beige and brown statue of the saint also stood in the old Port-au-Prince cathedral, before it was replaced with a smaller model in blue and red by that devoted scholar of applied Folkloristics, Dr. François Duvalier.[23] These and similar imported representations had affected, at least sartorially, even the heroes of the Revolution.[24] Rigaud claims Louverture covered his head with a red handkerchief, a sign that he "marched on the military points of *Le Mystère*, Nago Ogou Fer."[25] Dessalines, too, acknowledged Ogou as his *mèt tèt*, and publicly appropriated his attributes (red clothes, avid rum consumption, sword constantly at hand). In this guise Dessalines merged with Ogou in popular imagination, and so manifested himself at a ceremony described by Papa Doc:

> I quivered with stupefaction when the personality of the oungan was capsized in a hypnotic state, and surging forward from the depths of his consciousness came Dessalines the Emperor. It was truly him. The wild face, the fanatic physiognomy . . . he mounted two men in order to better arch his back in the pose of a chevalier.[26]

This manifestation seems entirely consistent with the representation of Papa Ogou as St. Jacques, seated on the back of his ounsis as he would mount his white horse, brandishing his sword to the right and to the left.[27]

But even more intense conflations of image, ritual, and history occur. Maya Deren observed and commented on an ounsi, possessed by Ogou, who was singing "I am wounded, oh I am wounded": "In this moment, with the side-stretched arms, the drooping head, the profoundly noble expression of the face, the attenuated, fallen posture of the body, and the tenderness of his two supporters, whose bodies are slightly bowed beneath his weight—he becomes the uncannily precise image of Christ being taken from the cross."[28] Aware of the extraordinarily complex borrowings suggested by this manifestation of Ogou as Jesus, Deren notes, "It is improbable that these people have anywhere seen that Christian image. Yet even if they had, and if this were an unconscious recreation of it, it would be the ultimate testimonial to their profound perception of the meaning of Christ. And if the image is not derived, but original to [Vodou], that also testifies, in another way, to an equal profundity."

But is there need for such Jungian hedging? What Vodouist does not also identify her religion as Catholic? And what Catholic Church is not adorned with the Stations of the Cross, including this very image of the Deposition—as either a frieze, or a lithograph? Such representations inspire a kind of mimetic frenzy in diverse post-Counter-Reformation cultures. In Haiti, they are objectified in the dramatis personae on the peristyle floor. In the Western World their influence persists in the campy half-worlds of Warhol or Mapplethorpe, whose art consists of recomposing the same lithographs that give a universal validity to the lwa in the street theology of Haiti.

It is impossible to analyze the transformations wrought upon the figure of Ogou in the last century without appreciating the profound influence of chromolithographs upon the religious imagination of Vodou.[29] All the major lwa are represented by these mass-produced, glossy images *signifying* (in the African American sense) correspondence between saint and lwa. Noting Vodou's ritual appropriation of chromos, scholars commonly dismiss the imported images as ruses used only to shield the true identity of a proscribed African god.[30] But ruses would hardly have become treasured commodities, sold in abundance, and at a relatively stiff price, to eager Haitian customers. Nor would ruses be so widely and intimately employed to adorn home and ounfò.

The chromo trade began in the mid-nineteenth century, after a concordant was signed with the Vatican ending a schism that began in 1804. The first imports from Europe were an immediate hit. Their continuing mass popularity confirms not just the imposition but the co-option of Catholicism. Replete with cabalistic imagery, chromolithographs have become rich food for an African religious sensibility cut off from its native sources. Far from being peripheral, chromolithographs constitute the single most important contemporary source for the elaboration of Ogou theology.

The bewildering array of folk exegesis inspired by the lithograph of St. James indicates some dimensions of the hermeneutic treasure house offered up to the imagination of the Vodouists. Leiris wrote the first and still most important essay on the subject:

> One sees St. James on horseback, with a sword and shield, fighting the infidels and escorted by a knight in armor . . . for all my informants, the main character is the god of the forge and war, Ogou Feray or Ogou Fer (who has a sword as his essential attribute, and as with the other Ogous, red as his color). But for some, the second character is Ogou Badagri, brother of Ogou Feray, even though for others, he is more likely a Gede, spirit of the cemeteries; that because of the lowered visor of the helmet of the character in question recalls the chin piece and other cadaverish aspects (such as cotton in the nostrils) with which the adepts who incarnate the Gedes frequently make themselves grotesque, all things which, besides, explicate why these latter are reputed to always speak through their noses.[31]

Note that the chromo inspires more than interpretation. It also inspires narratives about the deity and his relationships which constitute a living Vodou mythology. The chromo becomes a revelatory source, open to counter-analyses like rival Jewish, Christian, or Muslim hermeutics of the same Old Testament text. Thus Leiris's informants suggest alternative folk exegeses of the masked knight behind St. Jacques which identifies him not as Gede, but as Ogou Badagri. His visor has been lowered by his brother to prevent him from courting Ezili, whose favors both are seeking. So story becomes plot, and plot thickens.

Milo Marcelin identifies the masked figure as St. Philip, the twin brother of St. James, and so recognized as Ogou Badagri blinded by a brother's jealousy: "There is a battle between these Ogou warriors, proud, jealous, especially in those things which concern affairs of the heart. To whom does this apply but to the wanton Ezili? A question which can never be resolved, because no one could possess Ezili, but which gives place to a number of myths associated with Ogou and Ezili."[32] In Haiti, myths become reified. Note that the dramatis personae of this variant, Sts. Philip and James, are now emblazoned on a Port-au-Prince tap-tap, their celestial rivalry transformed into urban decor.

Leiris offers this cogent analysis of the chromo-hermeneutic process:

> Plurality of attributes and of names for the same divinity or the same saint (between which a very extended and very complex play of corresponding elements of identification could be operative), extreme elasticity of possibilities of rapport . . . variability of representations attached to the same divinity, and variability in the interpretation of forms, forces one to . . . pay attention to everything—for historic conjunctions, and social conditions are of a kind that favor a syncretic process.[33]

By this very interpretative process, General Ogou acquires his attending officers: St. Michael and St. George. Some of Leiris's informants recognized Ogou Badagri in the chromolithograph of St. Michael the Archangel wearing a red cloak. In his left hand he holds a balance, in his right a sword with which he gets ready to rip apart a demon. Ogou Balendjo is detected in St. George, an equestrian figure also dressed in a red mantle, armed with a lance and spearing a dragon. Leiris notes that both saints appear to be warriors *and* blacksmiths, the Ogou family trades. The presence of fire in both images (St. Michel encounters flames from hell, St. George braves the dragon's tongue) enforce this double identification with Ogou. Like Ogou's horses on the peristyle floor, Sts. Michael and George don't fear fire while walking on burning coals or handling bars of red hot iron.[34]

Given the breadth and fuzzy margins of the Catholic cult of the saints, and the iconic resourcefulness of printing houses in Mexico and Italy, Vodouists do not lack for material upon which to work their hermeneutics. Each new iconic interpretation is contingent on the attributes of the last, no official tradition powerful enough to check theological innovation. The process is centripetal, pushing out into new forms like a jazz riff. So during the long nightmare of the Cédras regime (1991–94), chromos of St. Elias became a seller in the Iron Market. Saber in hand, this saint was poised to slice off the head of an underfoot victim, while all around him flames arose, and other victims were being executed beneath the image of the Golden Calf. "Ogou Kriminèl," one informant confirmed. And so indeed St. Elie must have appeared to the masses absorbing the blows of the Cédras-FRAPH thugs: sons of General Ogou transformed into criminals.

One need not look far beyond the agility of this folk hermeneutic to understand the amazing intellectual and aesthetic creativity which generates the sacred arts of Vodou. Nor, by the same token, should one be surprised at how easily many spokespeople for contemporary Vodou adopt New Age jargon. The Godfather of this sort of Vodou is probably Milo Rigaud, whose *Secrets of Voodoo* offers the following gloss on the origins of Ogou:

> Ogou Bhathalah (Obatala) is a part of the army of Ogou loas considered in the African tradition as the fathers of alchemy. In the universal tradition, Bhathalah corresponds to the first blade of the Tarot (the Juggler), whose hat is the sign of the "universal life." This is the Magus (Wise One) or spirit par excellence. Bhathalah personifies the "discipline of chaos" because it is he who directs, with the magic wand, the cosmic traffic. Bhatahlah "disentangles the roads" by placing himself, like a traffic officer, at the magic

> crossroads. The scriptural reference to his occult function is found in Ezekiel XXI, 19–21: "The Lord says, Appoint thee two ways, that the sword of the king of Babylon may come. . . . For the king of Babylon stood at the parting of the way, at the head of the two ways, to use divination."
>
> The "swordstroke of Ogou" means that the Ogou mystères (loas of fire) or "stellar powers, creators of the intellect" are descended from heaven through a fissure shaped like the female sex organ represented by the planet Venus (Erzulie, in Voodoo). These are the mystères who are "forgers," like the aelohim which descend from the Jewish *Aziluth*, or the "Lords of the Flame" in Indian mythology. The Ogous in the Voodoo tradition bear the "fire of heaven" or the "luminous fire of Venus" shaped in the forge to represent a short serpent which traverses the planetary earth and sows fire. "Ogou's swordstroke" means also the act of kissing the top of the flagpoles during the course of the ritual Voodoo salutations performed by the flagbearers. In Voodoo, Luci-Fer whose name is also Ogou-Fer, is Venus, the morning star. It is called "Bayacou star" because it accompanies the sun each morning to explain the "earthly necessities" or "needs."[35]

It would be easy to dismiss this theosophic farrago for the bizarre conflation of myths which it is. All the hallmarks of the New Age philosophy are present: the assumption of a universal solar mythology, gnosis, alchemy, secret language. These are all "keys" which unlock the "hidden meaning" of Vodou, including the meaning of the lwa who are at some dark level manifestations of a universal pantheon. Thus in one short paragraph Ogou is conflated with Obatala (as he is later with Shango); the Juggler and Magus of the Tarot deck; Ezekiel's prophetic sword; the Jewish *aelohim*; the Hindu Lords of the Flame; the morning star; and through an etymological trick, with Lucifer, the fallen angel; not to mention others in the Vodou pantheon—Danbala the serpent divinity and Legba/Kalfou, magician of the crossroads.

It would be a serious mistake, however, to disregard Rigaud's metaphysics. He is an extremely influential source for contemporary Vodou mythology, especially as the population of Haiti urbanizes. There has been a long and complicated fascination with Euro-Semitic mythologies in New World African cultures, which leaves its enduring mark in the proliferation of Freemasonic and Rosicrucian temples. We have already discussed elements of Masonic art and ritual which grace the Vodou ceremony. A visit to Port-au-Prince Temple Sacre 5005 reveals the influence of these cabalistic riffs in wall murals of cyber-mythology. Other trips to other temples in the city affirm that Ogou mythology has been refracted through multiple lenses: the imagery of Roman Catholicism especially, but also the detritus of popular culture and the alchemy of spiritism. Haiti's Ogou is now caught between St. Jacques, Alan Kardec, and Sly Stallone.[36]

With the folk hermeneutic explored, it becomes easier to understand why the women in the church at Plaine-du-Nord confronted the image of St. Jacques like hungry lovers. And why they still revere his empty niche. By the same token, the motivations of the priests who emptied that niche, and locked the iron gratings around the church, are also clearer. The actions of all the dramatis personae at Plaine-du-Nord are motivated, to varying degrees, by a common appreciation that the image of god may be more immediately sacred than god himself. Vodouists, Catholics, and

Protestants, at some level, all have intuited the truth of Baudrillard's powerful observation:

> What becomes of divinity when it reveals itself in icons, when it is multiplied in simulacra? Do these images mask the platonic idea of God, or suggest that God himself has ever been his own simulacra—that the images concealed nothing at all. In fact they are not images such as the original model would have made them, but actually perfect simulacra forever radiant with their own fascination. . . . Then the whole system becomes weightless, it is no longer exchanging for what is real, but exchanging in itself, in an uninterrupted circuit without reference or circumference."[37]

Baudrillard describes the power of images to create their own self-referential world. That is the world Sen Jak rides towards in Rameau's painting, "The Vodou gods ponder Haiti's destiny." As if emerging from a chromolithograph, he leads his white steed towards the two Ezilis, Bawon, Zaka, and Danbala, seated on little peristyle chairs around the vèvè Milokan, which calls down all the divinities. His arrival will close the circle of an imagined heaven where the fate of Haiti will be played out by all the painted saints.

Kòk Kalite

Vodou's endless ability to recycle imagery affects secular as well as religious life, though in Haiti the two have always been very difficult to disentangle. There is no "was" in Haitian history. The heroes and events of the Revolution are still debated in the newspapers, and by illiterate peasants. The 1806 assassination of Jean-Jacques Dessalines is still hot news. Was he done in at Pont-Rouge by a plot of the mulatto elite? This question carries weight in a nation where class and color lines usually coincide. Papa Doc, the self-proclaimed champion of *noirisme*, elevated Dessalines to first place in the national pantheon. The main boulevard in Port-au-Prince was renamed after him, and the gray guinea fowl (*pentard*), which had been Dessalines's symbol, became the emblem of Duvalier.

Veneration of the pentard as the national emblem must have influenced the selection of the fighting rooster (*kòk kalite*) as Titid's political emblem. The rooster plays a powerful role in the Haitian culture. This feisty bird is combatant in the cock fights which are a masculine preoccupation in Haiti. A tough never-say-die "quality cock" can carry a man's pride, and his betting money, far in these bloody fights. And that combative toughness is just the image Titid projected on the streets. He is the *kòk kalite*. His place on the ballot was marked by that bird. All his campaign posters, and the T-shirts, key chains, visors, caps, and wall art they inspired, carried a picture of Titid juxtaposed to the red rooster. There were so many flocks of rooster masks during 1991 Carnival celebrations which immediately followed his inauguration, that the streets of Port-au-Prince resembled a convention of Kentucky Fried Chicken restauranteurs.

Often these popular representations showed the rooster pecking at, attacking, or mounting the pentard. A Carnival masquerader held a rooster wrapped in red and blue bunting in one hand, and dangled a wretched pentard by her tied feet from the other. Even one of the topical Carnival songs scorned the last provisional president in the language of the birds. The lyric accused her of supporting a pre-emptive coup attempt against Aristide:

Erta Pascal Trouillot
Mama Kaka
Look how you let the pentard
Get into the national coop

The cock carries further Haitian connotations. It is a Masonic symbol of fortitude. It is a sexual symbol, a French military insignia, and perhaps most significantly, a key player in a very popular chromo of the crucifixion. In that picture, found in many ounfòs, the cross is surrounded by various symbols of the passion, including the pillar where Christ was scourged. On top of the pillar is a red rooster. In Christian symbology he represents the cock that crowed thrice to mark Peter's betrayal. But in Vodou? Several Haitians pointed out that the red rooster is sacred to Ogou, and its position in the chromolithograph is evidence of Ogou's presence at the supreme drama of Calvary. One young Vodouist told me explicitly that Titid's *kòk kalite* proved he, too, was a worshiper of Ogou and a Vodouist. Didn't everyone see him kiss his mother on T.V. and wasn't she wearing a red dress—the color sacred to Ogoun? As if to prove his point, the young man then pointed to an Aristide campaign poster. Under Titid's slogan, and in place of his face, the poster simply carried the pillar and rooster from the chromolithograph. No need for any further explanation. The message was plain enough for Vodou.

To many, Titid's 1990 election victory on a slogan of "Justice, Participation, and Transparency" was a miracle. When Roger Lafontant, one of the Duvalier old-timers referred to as "dinosaurs," tried to pull off a postelection coup, the streets exploded. Mobs rooted out and sometimes murdered suspected *makouts* and dinosaurs. The coup failed. And in the days preceding Carnival 1991, Titid's election was celebrated by a Te Deum Mass. Inside the church the horrified elites looked on, while outside, ecstatic crowds pressed at the windows and swung from limbs of a concrete crucifix shouting "Yo sezi. Yo sezi"—"They are shocked. They are shocked"—the very same taunt seen atop the tap-tap dedicated to Sts. James and Philip.

The fairytale ended in September 1991. After only seven months the elites and their *makouts* could not tolerate justice, transparency, participation, paying taxes, and especially interference with their lucrative drug trade. Titid became another in a long line of Haitian presidents packed off into exile. But unlike those other exiles, more legends grew in his absence. Some said that at the moment the soldiers came to kill him, Titid was possessed by Ogou and his assassins backed away. Others said he was protected by Christian powers, recalling wall posters of Titid with Jesus or plastic fobs with his familiar face beaming from inside the Immaculate Heart of Mary.

Since most Haitians are Catholic *and* Vodou, all these legends rang true, including the extraordinary vision of a prostrate Aristide sprawled before the putschists in Duval-Carrié's "*La Voix des Sans Voix*" [The Voice of Those Without Voices]. In this painted tableau, Titid is separated from the generals by Death in the tragicomic form of *niambo*, the Kongo personification of the moral authority of the dead.[38] Gazing down from the wall are the heroes of the Revolution, none of them strangers to the terrible machinations of power spread out below. *Niambo* acts as their surrogate, an imposing reminder (like his Haitian counterpart Gede) that the ancestors are not indifferent to acts of injustice.

Another measure of the longing inspired by Titid's exile were the flags created by Oldof. After the coup he began to make banners in honor of the once-and-future president, transforming a genre hitherto dedicated exclusively to the lwa. The production and sale of these flags was secret, and often by commission. Oldof's designs incorporated extravagant "quality cocks" alongside the more usual Catholic imagery, an elevation in the aviary which fused Titid's *kòk kalite* with the sacred cocks of the Crucifixion and Ogou. In one such flag, a massive, surreal kòk is standing atop a tomb containing the crucified Christ. The tomb is banked by skulls, and by crossed Haitian flags. The juxtaposition evokes the correspondence of Titid, Christ, Ogou, and of course Gede, whose skulls represent not only death, but the promise of regeneration.

Oldof did not live to see the second coming. He died in August, 1994, three months before President "Mirak" returned to Haiti. With Titid's restoration, the production of Lavalas flags openly flourished, especially in the Bel-Air atelier of Silva Joseph. In Silva's *drapo*, references to Gede, Bawon, or Christ Crucified disappear. *Kòk kalite* now emerges multicolored, triumphant, even vainglorious, heralding the face of Titid in sequins and pearls. In Silva's flag, the president's image usurps the place otherwise reserved for chromolithographs of the saint/lwa.

As if to acknowledge this apotheosis, a portrait of President Aristide now rests on a Petwo altar in an ounfò near Solomon Market, just a few minutes away from the national palace. Oungan Sauveur St. Cyr nestled the picture there, near a ceramic figure of Ezili Dantò, sometime after Titid's triumphant return in October 1994. Ogou at last was back home. And back in confederacy with Ezili Dantò and the other Petwo spirits with whom he had marched in 1791. In Vodou, as they say, there is no Was.

Notes

1. These observations on the ceremonies at Plaine-du-Nord are derived from my first field visit to Haiti in 1986. That visit was funded by a research fellowship from the National Endowment for the Humanities, and by the University of California Research Expeditions Program (UREP), which also sponsored the participation of ten research assistants. I wish especially to acknowledge the collaboration of Project Co-Director, Henrietta Cosentino. In turn, we wish to acknowledge Fr. Keweillant, the town pastor, for obliging us in many ways during our festival research.
2. Dating the origins of the festival is problematic. A woman from Plaine-du-Nord reported to Tele-Haiti, "During Hurricane Flora in 1960 a pit appeared in an open space within the village. Later it filled up with rainwater and formed a mud pool. The pilgrims of St. Jacques saw a direct manifestation of the power of Ogou in this event" (Chantal Regnault, personal communication). Carole Devillers has established a time line for this festival by a much earlier dating for the first appearance of the Trou, "Popular memory has it that these holes

started in 1909 when the Gallois River flooded the newly built Centenary Road. Later the pond grew bigger when sand was dug from it for construction of the police station" (Devillers [1985] 404). Keweillant further muddies the mud by asserting that the *basin* and bull festival for Ogou didn't begin until the 1930s.

3. Weekly ceremonies are suspended during the month of November and the season of Lent. November is dedicated exclusively to services for Gede, whose very interesting relationship to Ogou has been explored by Karen McCarthy Brown, "The Vèvè of Haitian Vodou: A Structural Analysis of Visual Imagery," PhD dissertation, Temple University, 1976. During Lent all liturgical services for the lwa are suspended, and their altar objects are shrouded in cloth. Santeria and other African derived religions practice a similar custom, which was common in pre-Vatican II Catholicism. There is the further implication of hierarchical ordering in this covering: the lwa paying homage to Bondye, in his aspect as the *mystère*, Suffering Jesus.
4. For the ritual significance of *rad penitans*, see Thompson (1983), chapter 3. Just as at a Vodou ceremony (or indeed, at a Catholic Mass), there were many more women than men pilgrims. The ratio was approximately 3:1.
5. Keweillant's observations come from an interview at his rectory on July 24, 1986. He later allowed us into the church to video the barrage of sacred offerings through the closed gratings. It seemed like a war zone. We noted that the guards posted inside were willing to accept offerings for St. Jacques and place them in the church in return for a small gratuity.
6. Not that the West African antecedents of Ogou present themselves as models of unilineal descent. The degree to which the Yoruba Ogun may be comprehended as a single deity with a particular iconography is a subject of some scholarly debate. Thus Karen Barber observes, "Like other *orisa*, Ogun is distinct and yet not distinct, participating in a spectrum of and capabilities shared by the whole array of spiritual beings. . . . He exists in a complex shifting configuration of relationships, sometimes overlapping, sometimes separated, in some towns occupying one role, in others another" (1990, 290).
7. The shrine built for the "Thunderer" over the site of his relics in Compostela became Europe's second most important pilgrimage site (after Rome, of course). The Wyf of Bathe claims to have visited Compostela during the opening lines of her prologue in *The Canterbury Tales.*
8. Ogun's patronage of the road is also honored in Western Nigeria where some taxi drivers seek to appease the orisa by the propitiatory hitting and killing of a dog, Ogun's favorite sacrifice. For a celebrated development of this theme, see Wole Soyinka's play, *The Road*, or his volume of poems, *Idanre.*
9. Barra was also spurred into creativity by a more secular paper airplane he observed in the collection of Marilyn Houlberg (Houlberg is professor of art and anthropology at the School of the Art Institute of Chicago). It is Barra's peculiar talent to find inspiration in whatever pitches up at the market, or in his dreams (private communication with Houlberg, 1994).
10. From a review of "Pulp Fiction," by Anthony Lane, *The New Yorker*, October 10, 1994, 96.
11. Vicki Goldberg, from her review of the "Elvis and Marilyn" Show at the Boston Fine Arts Museum: "A Pair of Saints Who Refuse to Stay Dead," Arts and Leisure Section, *New York Times*, December 18, 1994, 49.
12. LeGrace Benson, "Islamic Motifs and Haitian Art," *Journal of Caribbean Studies*, 9, (1 and 2), 62.
13. Thompson, *Flash of the Spirit*, 172.
14. Balandier (1968[1969]), 46.
15. See ibid.; Thornton (1984) inter alia.
16. Balandier (1968 [1969]), 49. Students of oral tradition will note the parallels between this testimony and that which marked the apparitions during the *reconquista*, or indeed the

apparition of Constantine, "In Hoc Signo Vinces," the paradigm for all "Cross in the Sky" legends.

17. Balandier (ibid.) goes on to make an important point about Afonso's co-option of this Christian festival, "It was merely a way of adding a new celebration to an older one, without changing the meaning of the institution as such. . . . Christianity was conceived as a supplementary method of reinforcement, not as a religion exclusive of the old beliefs. From this period began the 'heresies' which were to multiply until the end of the modern colonial period."
18. Balandier (1968 [1969]), 119–120.
19. According to Thornton, devotions to the Saint became institutionalized in Kongo: "in 1622, King Pedro II made a charter for the funding of his royal chapel, the Chapel of St. James, stipulating exact salaries to be paid to members of the chapel" (Thornton [1984], 161).
20. Even the Cathedral in Port-au-Prince, dedicated in 1781 to Our Lady of the Assumption, was widely understood to be under his protection and popularly known by his name (Beauvoir-Dominique [1991], 47).
21. Quoted from *The Bulletin of Our Lady of Perpetual Help*, a periodical founded by Msgr. Kersuzan, Bishop of Cap Haitien at the turn of the 19th century, cited in Pean (1977) (my translation).
22. Amy Wilentz has also described the attempt to seal up the Trou, "At the end of February, Keweillant told me, a group of militant Protestants came to see him, and asked whether he would help them plug up the source of the mudhole. 'It was a good idea, I told them. But now I realize that they should have done it right after February 7, or they should have waited much longer and done it when the people were more psychologically prepared.' . . . The battle is not a new one. Protestants and the Catholic Church have always wanted to stop up the hole; in Duvalier's day, Keweillant says, 'the hole was protected by the palace,' meaning the dictatorship. Now, it is protected by the people." (Wilentz 1987, 120–121).
23. The cathedral statue was part of a side altar dedicated to St. Jacques, which in the 19th century became popularly associated with statues of the revolutionary heroes erected in the nearby Champs-de-Mars (Beauvoir-Dominique, *L'Ancienne Cathédrale de Port-au-Prince*, 47).
24. The dialectic works in two directions, as Beauvoir-Dominique observes (*L'Ancienne Cathédrale de Port-au-Prince*, 62), "We must guard against a unilateral vision of the relationship between tradition and history as one of agent to product. While the collective memory of Ogou-Sen Jak worked on the heroes, the actions of history worked on the memory" my translation.
25. 1969 [1985], 66.
26. Beauvoir-Dominique, *L'Ancienne Cathédrale de Port-au-Prince*, 59.
27. Ibid., 59–60.
28. In her own footnote, Deren remarks, "This manifestation occurred in precisely the same way several times, over a five month period, and was accepted without surprise, as if those present were familiar with it, although it does not seem to be a very common aspect" (1953, 132).
29. The most significant parallel in contemporary religions is the dialectical relationship between the Indian chromolithograph of the Snake Charmer and the development of Mami Wata cults in West and Central Africa (Drewal 1988).
30. Included among the scoffers is Luc de Heusch, who dismisses the correspondence between lwa and lithograph, "But at the outset, we must do away with the fake problem of Christian syncretism in voodoo. Throughout Haiti, the temples are decorated with colour

reproductions of Catholic saints that evoke the loa through what Michel Leiris has referred to as a pun based on objects instead of words (*un calembour d'objets*). For example, Ogun the Yoruba war god, is likened to St. James portrayed as a victorious knight wielding a sword. Ogun belongs to the Nago family of gods, the Fon term for their Yoruba enemies. An especially well-qualified informant stated that the Nago spend their time making war on horseback. Hence Ogun, whenever he enters the bodies of his followers, is like a knight on horseback wielding a sword. Thus is explained the iconographic equivalence between the African god and the Catholic saint" ("Kongo in Haiti"), (291–292). De Heusch then makes the following definitive observation, "Any religion, including Christianity, is ultimately a syncretic phenomenon" (298–299). What DeHeusch fails to understand is that syncretism, once begun, takes on a life of its own.

31. Leiris 1952, 204 (my translation).
32. Marcelin quoted in ibid., 1952, 204, n. 2 (my translation). Comments on Marcelin are also included from Beauvoir-Dominique, *L'Ancienne Cathédrale de Port-au-Prince.* Comparative mythologists will note that Ezili fills the mythic place held by Osun in Yoruba/Lucumí theology. Both deities remain unattainable objects of Ogun/Ogou's frustrated libido; a frustration which may have inspired the fervent sexual offerings to St. Jacques reported by Fr. Keweillant at Plaine-du-Nord (see above).
33. Leiris, "Notes sur l'usage de chromolithographies catholiques par les vodouisants d'Haiti," 204, n. 3.
35. Rigaud 1969 [1985], 76–77.
36. His modern predicament is elegantly captured by Hector Hyppolite in his painting, "Ogou Ferraille." Gone are the martial clothes of St. Jacques. Hyppolite's Ogou is a Magus, controlling elements of divination and magic suasion as diverse as the catalogue of Milo Rigaud. See Drot, *Art Naïf, Art Vodou* (1988), 129. It is worth noting that in African cosmologies the forge is widely associated with mysteries of divination and magic.
37. Baudrillard 1984, 255.
38. A *niambo* is "a massive 'mummy' meant to be displayed, wept over by descendants, and then perambulated on in glory to its grave. . . . These towering, massive, gigantic presences, larger than life, loom over the person escorting them to their final resting place" (Thompson and Cornet 1981, 58). Of niambo's moral authority, Fu-Kiau states, "When you die you automatically become an ancestor. But not everyone is made into a niambo. Being buried in a niambo figure means the community believes this person will become our medium . . . saying to our ancestors, 'I am their mediator. I move between worlds' Thompson, 63. So in Duval-Carrié's painting, niambo stands between Titid, Cedras and *les héros*, mediating the unfolding disaster for Haiti.

References

Balandier, Georges. 1968 [1969]. *Daily Life in the Kingdom of the Kongo.* London: George Allen and Unwin Ltd. [Cleveland: Meridian].

Barber, Karen. 1990. Review of "Africa's Ogun." *Journal of Religion in Africa* 20: 3.

Baudrillard, Jean. 1968. *Le Système des Objets.* Paris: Gallimard.

———. 1984. The precession of Simulacra. In *Art After Modernism,* edited by Brian Wallis. New York: The New Museum of Contemporary Art.

Beauvoir-Dominique, Rachel. 1991 *L'Ancienne Cathédrale de Port-au-Prince.* Port-au-Prince: Editions Henri Deschamps.

Benson, LeGrace. 1992. Some observations on West African Islamic motifs and Haitian art. *Journal of Caribbean Studies* 9 (Winter/Spring): 59–66.
Brown, Karen McCarthy. 1976. The Vèvè of Haitian Vodou: A structural analysis of visual imagery. PhD dissertation, Temple University.
De Heusch, Luc. 1989. "Kongo in Haiti: A New Approach to Religious Syncretism." *Man* (Journal of the Royal Anthropological Institute), 24 (June): 290–303.
Deren, Maya. 1953 [1970, 1983]. *Divine Horsemen: The Living Gods of Haiti.* New York: Chelsea House; McPherson and Co. and Thames and Hudson.
Devillers, Carole. 1985. "Of Spirits and Saints," *National Geographic* 167 (3).
Drewal, Henry John. 1988. "Interpretation, invention, and representation of Mami Wata," *Performance in Contemporary African Arts.* Bloomington: Indiana University Folklore Institute.
Leiris, Michel. 1952. "Notes sur l'usage de chromolithographies catholiques par les vodouisants d'Haiti," *Les Afro-Americans.* Dakar: IFAN.
Pean, Marc. 1977. *L'Illusion Héroïque: Tome 1 1890–1902, 25 Years of Capoise Life.* Port-au-Prince: Prix Deschamps.
Rigaud, Milo. 1969 [1985]. *Secrets of Voodoo.* San Francisco: City Lights Books.
Soyinka, Wole. 1967. *Idanre and Other Poems.* New York: Hill and Wang.
Thompson, Robert Farris, and Joseph Cornet. 1981. *The Four Moments of the Sun.* Washington, DC: National Gallery of Art.
Thompson, Robert Farris. 1983. *Flash of the Spirit: African and Afro-American Art and Philosophy.* New York: Random House.
Thornton, John. 1984. The Development of an African Catholic Church in the Kingdom of the Kongo, 1491–1750. *Journal of African History* 25: 147–167.
Wilentz, Amy. 1987. Voodoo in Haiti today. *Grand Street* (Winter).

Notes on Contributors

Patrick Bellegarde-Smith is Professor of Africology at the University of Wisconsin-Milwaukee. He holds a PhD in International Politics and has written extensively on issues of Haitian cultural identity, social thought and religion. He is a *houngan asogwe*, a priest of Vodou. He is the author of *Haiti: The Breached Citadel* (1990 2004), *In the Shadow of Powers: Dantès Bellegarde in Haitian Social Thought* (1985), and editor of *Fragments of Bone: Neo-African Religions in a New World* (2005). He is co-editor, with Claudine Michel, of the forthcoming anthologies *God in Every Woman: Gender, Power and Politics in Haitian Vodou* (forthcoming) and *Haitian Vodou: Spirit, Myth, Reality* (Indiana University Press, 2006).

LeGrace Benson holds a PhD from Cornell University and an M.F.A. from the University of Georgia. Currently she is Director of the Arts of Haiti Research Project and an Associate Editor of the *Journal of Haitian Studies*. Author of a number of articles in scholarly journals concerning Haitian art, she has also contributed chapters to books concerning educational, environmental, and art issues in Haiti and the wider Caribbean. She is now a visiting researcher at the UCSB Center for Black Studies.

Karen McCarthy Brown is Professor of Sociology and Anthropology at Drew University. She holds an MA from Columbia University and a PhD from Temple University. Her areas of expertise include religion and politics in Haiti and religion as a form of social resistance. She is the author of the acclaimed *Mama Lola: A Vodou Priestess in Brooklyn*, (1991/2004) and another important volume, *Tracing the Spirit: Ethnographic Essays on Haitian Art*, as well as many articles and essays.

Carrol F. Coates is Professor of French & Comparative Literature at Binghamton University (SUNY), where his teaching has focused on francophone literature of the Caribbean and West Africa. He has published English translations of three Haitian novels with the University of Virginia Press: René Depestre, *The Festival of the Greasy Pole*; Jacques Stephen Alexis, *General Sun My Brother*, and *In The Flicker of an Eyelid* (the latter in collaboration with Edwidge Danticat). He is Series Editor of Caraf Books (Virginia), Caribbean and African Literature in Translation.

Leslie G. Desmangles is Professor of Religion and International Studies at Trinity College in Hartford, Connecticut and served as the first president of the Haitian Studies Association from 1994–1998. He holds a PhD in Anthropology of Religion from Temple University. He has published many articles on Haitian Vodou and is also the author of *The Faces of the Gods: Vodou and Roman Catholicism in Haiti*

(University of North Carolina Press, 1993). He edited a volume for the Haitian Studies Association entitled *Haiti in the Global Context* (1995), and served as an associate Editor for the *Encyclopedia of African and African American Religions* (Routledge Press, 2001).

Gerdès Fleurant holds a PhD in music and is Professor Emeritus of Music Culture at Wellesley College. He is the author of many articles published in scholarly books and journals as well as *Dancing Spirit: Rhythms and Rituals of Haitian Vodun, the Rada Rite* (1996). He is the founder and director of Léocardie and Alexandre Kenscoff Cultural Center, a culture, research, and development center in Mirebalais, Haiti.

Laënnec Hurbon was born in Jacmel, Haiti and holds graduate degrees in theology (Institut catholique de Paris) and sociology (Sorbonne). He is the Director of research at the Centre National de la Recherche Scientifique in Paris (CNRS) and a Professor at Quisqueya University in Port-au-Prince, of which he is a founding member. He specializes in Caribbean religion, culture and politics, and is the author of many books and essays on Haitian Vodou, including *Dieu dans le vaudou Haitien, Voodoo: Search for the Spirit, Voodoo: Truth and Fantasy* and *Los Misterios del Vudu.*

Elizabeth McAlister holds an MA in African American Studies, an MA in History, and a PhD in American Studies. She is Associate Professor of Religion and Chair of the Department of Religion at Wesleyan University. Her area of expertise is Afro-Caribbean religious traditions, particularly Haitian Vodou. She is also interested in issues of transnationalism, religion and the social construction of race and ethnicity, as well as religion and gender and sexuality. She is the author of *Rara: Vodou, Power, and Performance in Haiti and its Diaspora* (University of California Press, 2002) and she has produced two CDs, *Angels in the Mirror: Voodoo Music of Haiti*, and *Rhythms of Rapture: Sacred Musics of Haitian Vodou*, published by the Smithsonian in conjunction with the vodou exhibit at the UCLA Fowler Museum of Cultural History titled "Sacred Arts of Haitian Vodou."

Claudine Michel is Professor of Black Studies at the University of California, Santa Barbara and holds a PhD in International Education. She is the author of *Offerings: Continuity and Transformation in Haitian Vodou* (forthcoming from Oxford University Press), *Aspects Educatifs et Moraux du Vodou Haïtien* (1995), and co-author of *Théories du Développement de l'Enfant: Etudes Comparatives* (1994). She is also co-editor of *Black Studies: Current Issues, Enduring Questions* (2001) and *The Black Studies Reader* (2004). She is co-editor, with Patrick Bellegarde-Smith, of the forthcoming anthologies *God in Every Woman: Gender, Power and Politics in Haitian Vodou*, and *Haitian Vodou: Spirit, Myth, Reality* (Indiana University Press, 2006). She is currently the Editor of *The Journal of Haitian Studies.*

Pierre Minn received his BA from Yale University. He is presently completing PhD studies in medical anthropology and related fields at McGill University in Canada. He is the recipient of the Peter K. New Award (1998), and has received a Fulbright Grant to South Korea. He has done field research on medical anthropology and religion in various parts of Haiti and the Dominican Republic.

Richard Brent Turner is Associate Professor of African-American World Studies and Religious Studies at the University of Iowa. He is the author of numerous articles, including "Mardi Gras Indians and Second Lines, Sequin Artists and Rara Bands: Street Festivals and Performances in New Orleans and Haiti," and "The Haiti-New Orleans Vodou Connection: Zora Neale Hurston As Initiate Observer," both published in the *Journal of Haitian Studies*. His most recent book is entitled *Islam in the African-American Experience* and his current research is on Vodou in New Orleans.

Anna Wexler is an artist and scholar, with a PhD from Harvard University, whose research/writings have focused on visual forms in Haitian Vodou and Cuban Santeria. Her multimedia performance piece, "Gullah Jack's Bag," was produced in collaboration with the artist/Vodou priest, Erol Josue, in 2002. She is currently developing collaborative performance projects engaged with persona, duration, and the production of ritual objects. A faculty member at the Springfield College School of Human Services, Boston, she teaches courses on African Atlantic and other culturally based healing systems.

Index

FSC
www.fsc.org
MIX
Papier | Fördert
gute Waldnutzung
FSC® C083411

Kolibri 360 GmbH
Ferdinand-Jühlke-Straße 7
99095 Erfurt, Deutschland
produktsicherheit@kolibri360.de